# Home Networking Technologies and Standards

For a listing of recent titles in the *Artech House Telecommunications Library*, turn to the back of this book.

# Home Networking Technologies and Standards

Theodore B. Zahariadis

Artech House, Inc.
Boston • London
www.artechhouse.com

Library of Congress Cataloging-in-Publication Data

Library of Congress CIP information is available on request.

British Library Cataloguing in Publication Data
Zahariadis, Theodore B.
    Home networking technologies and standards.—(Artech House telecommunications library)
    1. Home computer networks   2. Broadband communications system—Evaluation
    3. Broadband communication systems—Standards
    I. Title
    621.3'821

    ISBN   1-58053-648-4

Cover design by Christina Stone

© 2003 ARTECH HOUSE, INC.
685 Canton Street
Norwood, MA 02062

International Standard Book Number: 1-58053-648-4
A Library of Congress Catalog Card Number is available on request

10 9 8 7 6 5 4 3 2 1

*To Marianthi*

# Contents

**CHAPTER 9**

Firewire                                                      115

**CHAPTER 10**

Middleware Technologies                                       125

# Preface

Connecting each house to broadband access networks represents a great opportunity to offer added-value services and broadband Internet access to residential users, exponentially increasing the network operators' and service providers' customer base and revenues. Until recently, the major obstacle to the "digital networked house" was the inadequate access network infrastructure and the cost of new installations. Today, advances and innovations in broadband access technology have enabled a number of competing network technologies, ranging from copper enhancements to fiber, wireless, and satellite communication alternatives, while huge investments have brought the information superhighway just outside of a critical percentage of houses worldwide.

Despite the costly investments, the market of home networking services is evolving very tenuously. The new barriers are the insufficient, and in many cases proprietary home network, and the lack of a universal, widely accepted standard. Instead, there is a large variety of consortia and authorized standardization bodies that have been writing independent, and in many cases noninteroperable specifications, for residential networks. Dozens of candidate wireless and wireline technologies over multiple existing and/or emerging network architectures and physical media, aim to provide multimedia home systems (inter)communication.

This book aims to open the already ajar door of residential networking by providing a concise, though accurate presentation and analysis of the most mature and emerging technologies, standards and trends in the home networking arena to a large audience that ranges from academia to service and network providers. It aims to be a rapid introductory tutorial and reference book for the competing broadband in-home network technologies and standards.

# Acknowledgments

I would like to thank a number of people who worked to make this book a reality. First of all, I would like to thank Artech House for having foreseen the need for a book that would gather, compile, and analyze the current and emerging home-network standards. In particular, I wish to thank Dr. Julie Lancashire, senior commissioning editor, who accepted my proposal to write this book and gave me much advice on how to proceed, and Louise Skelding, editorial assistant, who looked after all details of the book and organized the day-by-day communication.

A special thanks to Dr. Nicholas Zervos, director of Ellemedia Technologies, who shared with me his vision of home networking and also contributed his world-renowned knowledge of transmission systems to the book, as well as to my R&D group at Ellemedia Technologies, who have worked intensely to realize the first RGs products. Also, a special thanks to Dr. Victor Lawrence, vice president of Bell Labs–Advanced Technologies, who has supported me in many ways and inspired pioneering work in the home-networking arena.

This book integrates knowledge from many sources, conference, magazine, and journal articles; books, standards, and alliances; technical and white papers; and Internet sites. I would like to thank each author, each company, and each organization that has contributed to home-networking research, because this book is based on the results of their efforts. A list of references used in each chapter is presented at the end of that chapter.

I would particularly like to thank Nick Hunn at TDK Systems for allowing me to use the company's market studies, and also Dr. Chatschik Bisdikian at IBM T. J. Watson Research Center, Dr. Peter Marshall at Philips, Professor David Wagner at Berkeley, and Dr. Alan Woolhouse, vice president of CSR.

Most of all, I would like to thank my wife, Marianthi, for her patience and support while I completed this book.

# CHAPTER 1

# The Magic Boxes

## 1.1 Introduction

I wish I could watch cinema from my home!

Television was invented to fulfill this wish approximately 50 years ago. The TV was a magic box with incredible capabilities and features, which was soon installed in each and every house—a magic box that could exhibit people, animals, birds, vehicles, airplanes, continents, civilizations, the whole Earth and all of history, space and the distant planets! Today TV offers amusement, awareness, education, and entertainment; however, it has lost its mystery and magic. It is just another consumer-electronic device, a box installed in the living room, the bedroom, or the kitchen, a box for passive users.

I wish I could select what I'll watch in my TV set and watch it whenever I want to!

This was the second step: the wish to control the TV content. The result was another magic box called the *video cassette recorder* (VCR). Millions of magic VCRs have been sold worldwide, thousands of films have been recorded to tapes, and thousands of video clubs sell or rent VCR tapes. VCRs offer the freedom to select the film or record a TV program and watch it whenever the user wants to. However, VCRs can only repeat a recorded movie, film, or program. Today, new wishes, new facilities, new "needs" have appeared.

I wish I could select the film or the program that I want to watch, just before I watch it, no matter whether it is afternoon, midnight, or a holiday, without having to move from my couch.

I wish I could watch the episode of my favorite TV program, which I did not have the chance to record, owing to a traffic jam on the way back from work.

I wish I could read my corporate e-mail from my kitchen during breakfast, while my wife is having a videophone conference in the living room with her sister in New Jersey, my daughter is speaking on the phone with a friend in Madrid, and my son is playing an interactive video game in his room with his cousins, who are located in Athens, and some friends, who are located in Sidney.

I wish I could control my home's security system and monitor the babysitter's behavior while I'm at work.

I wish I could communicate from my mobile personal digital assistant (PDA) with my house automation system while driving home and turn on the heating system, the coffee machine, and the water heater.

I wish my refrigerator could keep track of the stored goodies and automatically place all weekly grocery orders with the supermarket!

Some years ago these wishes could be the script from a science fiction film. Today, the only question is when and how all these wishes will be economically viable and available to consumers worldwide. The cornerstone for turning this wishful thinking into reality is again a magic box called the *residential gateway* (RG). The RG is a network device that interconnects the home PCs, controls and supervises consumer-electronic devices, and provides broadband home access. However, the home communications redefinition is more an evolution than a miracle.

First of all, it is the access network, which is the network that connects the house with state-of-the art, integrated data, voice, and video metropolitan networks. Advances in transmission and compression technologies, pioneering innovations, and huge investments have assured broadband and low-cost home connectivity [1]. Then, follows the intelligence of the home consumer-electronic devices. More and more houses are equipped with at least one PC, modern home security and automation systems, and network-aware devices. All of these devices may be interconnected via existing and future, wired and wireless, voice and data home networks. Finally, the RG is the magic wand, which turns this heterogeneous collection of devices into an intelligent, cooperating digital home.

However, digital home networking is not a fairy tale. It is a potentially huge, multitrillion-dollar market. This market is not restricted to new intelligent appliances or enhanced network devices. Connecting each house to broadband access networks represents an unprecedented opportunity to set the ground for a vast range of new home applications and value-added services to residential users and to expand the customer base beyond the saturated corporate environment.

## 1.2  Book Scope and Structure

Despite its promising prospects, the home-networking market is evolving very tenuously. One reason is the large variety of consortia and authorized standardization bodies that have been writing independent, in many cases noninteroperable specifications for residential networks. In the arena of indoor networking (Figure 1.1), more than 50 candidate technologies over multiple existing or emerging network

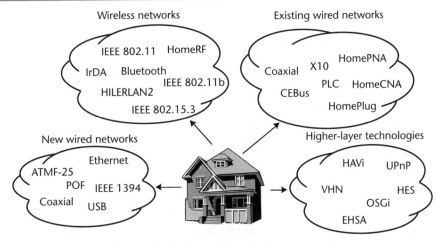

**Figure 1.1**    Home-networking technologies and standards.

architectures and physical media aim to provide multimedia home-systems (inter) communication [2].

This book aims to describe the candidate home-networking technologies by providing a concise, accurate presentation and analysis of the most mature and emerging technologies, standards, and trends in the home-networking area. It aims to be a brief introductory tutorial and reference book for the competing broadband in-home-networking technologies and standards.

The book is structured as follows: Chapter 2 describes the end-to-end reference architecture, where the main residential services are identified and their network requirements are briefly analyzed. It reveals the home reference architecture and places the home network into an end-to-end network/service architecture.

Chapters 3 and 4 describe the access network technologies. As the access network connects the home with the information superhighway and the Internet, the characteristics and capabilities of competing networking technologies strongly affect the in-home network's capabilities. Chapter 3 describes in greater detail the wired (wireline) network technologies, while Chapter 4 presents the fixed-wireless and satellite alternatives.

Then, the book tackles the home-networking technologies. It categorizes the available technologies and standards into three groups, based on the physical requirements of the medium.

Chapter 5 describes the technologies that do not need rewiring of the house and are based on the existing in-home structured wiring and cabling system. Current and future technologies that use the powerline and the phone-line home wiring, along with home cable networks wherever available, are described.

Chapters 6 and 7 describe existing and emerging in-home wireless technologies. The "no wires" technologies are considered the Holy Grail of home networking and are expected to play a key role in pushing forward the wide acceptance of the digital house. These chapters describe the wireless technologies that are expected to capture the maximum share of the home-networking market for different applications.

Chapter 8 presents the technologies that require new structured wiring. The importance of these technologies is based on their ability to provide a secure way to deploy new services, as they have been tested in enterprises and business sectors and are now brought to the home environment. Among various interfaces and technologies that fall into this category are Ethernet and the universal serial bus (USB). In Chapter 9, special focus is given to IEEE 1394 (Firewire) technology, as it is one of the most promising digital broadband, wired, in-home technologies.

Chapter 10 presents the scope, potential applications, philosophy, operational concepts, architecture, and protocol stack of higher-layer technologies that are independent of media and physical/transmission techniques and that aim to provide convergence between multiple in-home and access networks.

The RG topic, as a single point of networks convergence, is introduced in Chapter 11. Based on prototype development, laboratory experiments, and field trials throughout Europe, a modular and compact RG hardware architecture is described, while the software architecture to support the RG functionality is described and its main building blocks are analyzed.

Finally, Chapter 12 recapitulates the book content and provides some considerations about the future digital smart-networked home.

## References

[1]    Zahariadis, T., et al., "Interactive Multimedia Services to Residential Users," *IEEE Communications Magazine,* Vol. 35, No. 6, June 1997, pp. 61–68.

[2]    Zahariadis, T., Pramataris, K., and Zervos, N., "A Comparison of Competing Broadband In-Home Technologies," *IEE Electronics and Communications Engineering Journal (ECEJ),* August 200b2, pp. 133–142.

# CHAPTER 2

# End-to-End Reference Architecture

## 2.1 Introduction

Recent advances in signal processing and huge investments in access networks have brought the information superhighway to a large percentage of houses worldwide. More and more homes are already equipped with at least one PC or smart consumer-electronic devices, and the number of houses with an Internet connection is rapidly increasing. However, all these intelligent, and in most cases, network-enabled devices are isolated. Deployment of home networks and convergence with the broadband access networks represent an unprecedented opportunity to offer value-added services and broadband Internet access to residential users and to expand its customer base beyond the saturated corporate environment.

In this chapter we will identify the potential residential user services and service requirements. This route will smoothly reveal the home reference architecture and place the home network into an end-to-end network/service architecture. Moreover, it will provide a brief overview of the alternative and most competitive access network alternatives.

## 2.2 Residential Services Identification

Today, television, telephony, and radio are the most widely accepted services offered to residential users worldwide. It is difficult to make precise predictions about which systems, solutions, and services will be considered successful after 10 years. An important factor to this uncertainty is the limited deployment of broadband home networks and the fact that trials of residential broadband services have shown that what consumers find most valuable are Internet, Intranet, and e-mail access [1]. Although there is no widely recognized "killer app" that will convince reluctant consumers of the value of broadband connectivity, interactive streaming multimedia and information delivery services have already started to expand to the residential market and the demand for such services is projected to grow steadily [2].

A grouping of future home services is shown in Figure 2.1. Services expected to capture the greatest residential users market segments include the following:

5

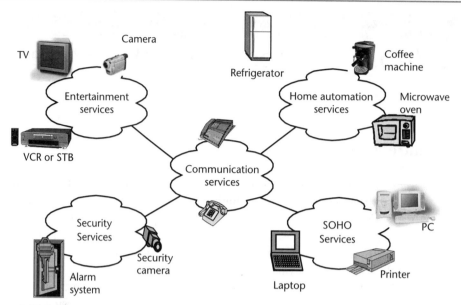

**Figure 2.1** Future home services grouping.

- *Home Communication Services:* Evolution of today's narrowband communication services is expected to be among the first service segments to develop. Future communication services will include applications like shared Internet access from multiple home PCs and Internet appliances, and value-added voice services, such as voice over IP (VoIP), videophone, and multiple phone connections over one line.

- *Small Office Home Office (SOHO) Services:* The number of telecommuters working from small offices and home offices is increasing rapidly. The nature of office work has changed; the time demands and cost of travel to and from work, especially in large cities, and the cost-saving measures taken by many companies have driven many employees to work from a SOHO environment. Additionally, increasing requirements and office responsibilities drive higher-level employees and managers to bring work home. To work efficiently from home, telecommuters must establish a SOHO network, access the corporate Intranet properly and securely, and establish cost-effective voice communications.

- *Home Entertainment Services:* Since 1994, video on demand (VoD)–like services have been expected to generate a telecommunications revolution and drastically increase the telecom operators and service providers' revenues. Expensive and time-consuming trials have taken place, but VoD has not proven itself able to capture the projected attention from residential customers. Some pollsters believe that the content of the trial films was not appropriate, others that the applications were not user-friendly enough, and others that the customers were just not ready for such a service. Most of them agree, however, that sooner or latter, entertainment services that include video streaming (e.g., multiplayer network games) will be widely deployed.

- *Home Automation Services:* Remote and unified control of smart, network-enabled, consumer-electronic devices and appliances will be a key service for future home networks. The scenario foresees the ability to connect to your home network remotely (e.g., while at the office or on the way home) and turn on the house heating system, the waterheater, or the coffee machine or start recording your favorite film on your home VCR. More trendy applications request intelligent appliances that initiate actions (e.g., the refrigerator does the weekly shopping from a selected supermarket via an automatic e-mail).

- *Home Security Services:* Alarm systems are already a relatively mature market. However, the convergence of security systems with home and access networks, along with modern monitoring and sensing devices (e.g., motion detectors, low-cost cameras), represents a new market potential. Subscribers will be able to monitor their homes remotely via Internet browsers or mobile PDA, or receive automatic e-mail if something happens. Applications may include not only alarms and security systems, but also babysitting services or health-care services for elderly and disabled people.

## 2.3   Services Requirements

Home services that will be considered successful after 5 or 10 years may differ from those that we identified in the previous section. However, their major requirements are not expected to differ significantly, while simplicity and affordability will be the top priority [3]. Projected future services will include the following requirements:

- *Bandwidth Requirements:* Future applications will include one or more video streams. Especially for entertainment services, high-quality video will be required. Table 2.1 shows the characteristics of the most widely accepted digital encoding standards.

  MPEG-1, MPEG-2, and MPEG-4 are ISO/IEC standards developed by the Moving Picture Experts Group (MPEG). MPEG-1 and MPEG-2 are the Emmy Award–winning standards that made interactive video on CD-ROM and digital television possible. As the table shows, MPEG-4 can achieve good to very good quality video at low data rates. However, most digital video films have already been encoded in MPEG-2, and transcoding to MPEG-4 is as expensive and complicated as transferring each VCR tape to DVD. Moreover,

**Table 2.1**  MPEG Video Encoding Standards Characteristics

|  | MPEG-1 | MPEG-2 | MPEG-4 |
|---|---|---|---|
| Standard available since | 1992 | 1995 | 1999 |
| Default video resolution (PAL) | $352 \times 288$ | $720 \times 576$ | $720 \times 576$ |
| Default video resolution (NTSC) | $352 \times 288$ | $640 \times 480$ | $640 \times 480$ |
| Maximum audio frequency range | 48 KHz | 96 KHz | 96 KHz |
| Maximum number of audio channels | 2 | 8 | 8 |
| Maximum data rate (Mbps) | 3 | 80 | 5–10 |
| Regular data rate (Mbps) | 1.5 | 6 | 0.9 |
| Video quality | Satisfactory | Very good | (Very) good |
| Encoding requirements | Low | High | Very high |
| Decoding requirements | Very low | Medium | High |

MPEG-4 encoding and decoding systems are not so popular yet, as their hardware requirements are quite high. On the other hand, MPEG-2 requires approximately 6 Mbps per video stream; thus, 6 Mbps can be considered the lowest bandwidth requirement for video distribution.

- *Quality of Service Requirements:* Besides bandwidth requirements, home networks should provide guaranteed quality of service (QoS), especially for streaming applications. Isochronous traffic, limited delay, and minimum jitter are among the requirements for multimedia applications like high-definition television (HDTV), Digital VoD, virtual reality games, along with traditional voice (telephone) services.

- *User Friendliness and Reliability Requirements:* Home-networking products and solutions target a large audience of customers with no computer experience. Thus, the technologies and the applications should be as consumer friendly as possible. Moreover, home-networking solutions should be easy to install, providing plug 'n' play (PnP) and/or autoconfiguration features, and should enable remote maintenance from the service, network, or manufacturer site. Reliability and robustness are also considered mandatory, as residential users will have difficulty identifying and handling problems and home-networking products need to operate all day (and night) long.

- *Reasonable Cost Requirement:* At the end of the day, services are measured in money. Thus, home-networking solutions should be provided at reasonable cost, comparable to home appliances like the TV and the microwave oven.

- *Low Installation-Cost Requirement:* Apart from the equipment cost, home-networking solutions should take care of installation costs. A 500-ft Ethernet cable may be enough to wire an average house; however, people may not will to drill holes into their walls to install a home network. Thus, wireless technologies or technologies that reuse existing wiring infrastructure of the home (e.g., phone lines, power lines, coaxial TV) may be preferred, at least for existing households.

- *Standards and Interoperability Requirements:* It is expected that a multitude of home-networking products, technologies, and solutions from different vendors will coexist in future homes. Having identified the interoperability challenge, many industry associations, consortiums, and working groups have already been formed to define appropriate open standards and conformance tests.

## 2.4   Home Network Reference Architecture

The provisioning of broadband networking services, including emerging, interactive, streaming multimedia services to residential users, requires a comprehensive broadband end-to-end digital network infrastructure spread from the service provider to the customer premises.

In the general scenario, a simplified reference architecture similar to the one shown in Figure 2.2 can be considered. The architecture is organized into three segments: the content or service provider, the central office, and the home network. The content or service provider segment is responsible for preparing, storing, and manipulating the multimedia content. Videos are received mainly in analog format from various sources: satellite connections, analog wireline or wireless broadcasting

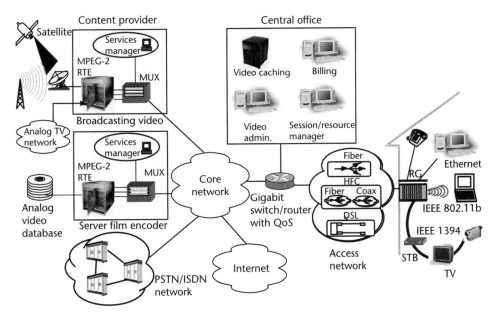

**Figure 2.2**   End-to-end reference architecture.

networks, or cable TV (CATV) networks. The initial procedure is to decrypt or descramble the received analog video. Compressed digital video storage and distribution is much more efficient in terms of size and bandwidth requirements, as compared with analog. Thus, analog streams are converted to digital MPEG-1, MPEG-2, or MPEG-4 format via carrier-class MPEG real-time video encoders (RTE). Subsequently, the digital streams are multiplexed over Internet Protocol (IP) and/or encapsulated into ATM cells via a multiplexer (MUX) and broadcasted or distributed over the core network. In this way, digital video and delayed video broadcasting applications can be provided, while a copy can be stored at the video server farm for on-demand retrieval applications. In all cases a service manager host synchronizes the video encoding, multiplexing, and distribution process.

The digital video servers in the video server farms are capable of simultaneously supporting thousands of independent, active multimedia sessions and storing enormous amounts of multimedia data. Each time a VoD or Near VoD service request is initiated, digital content from the server farm is retrieved, and a session between a server and the customer is set up via the load management switch. The switch balances the requests between the video servers and collects statistical information that can be used for billing the services and measuring the film's popularity. The latter can be used to improve distribution of film copies in network caches. Input for VoD-like services can also be retrieved from analog video databases and encoded in real time from the MPEG encoders.

We also assume that the traditional telephony network is located at the service provider side, although practically, it can be physically extended to the customer premises.

The multimedia content is distributed over the core network to the central office. According to the adopted network technology, the core network may use IP, Asynchronous Transfer Mode (ATM), or both protocols. At the central office, various servers are considered:

- A video administration server, which provides digital broadcast channel (BCH) attributes, channel permissions, and pay-per-view events, and logs transactions, and so forth;
- A session and resource manager responsible for setting end-to-end sessions and allocating the required network resources;
- A billing system that provides charging and billing information;
- A video cache server for temporal caching of the most popular films.

Access to the Internet is also considered a de facto service. Next generation broadband Internet may also be used for applications like IP telephony, video conference, or intranet access.

The interconnection between the core network, the central office, and the access network is performed by QoS-enabled gigabit switches or routers, according to the adopted transport protocol. Potential access networks include variations of digital subscriber line (xDSL), hybrid fiber-coaxial (HFC), fiber networks, and fixed-wireless technologies. According to supported services, a narrowband access network (e.g., plain telephony) may also be traditionally available.

Finally, at the user side, customer premises equipment (CPE) is required to decode and decompress the digital signal and handle upstream communication. Digital set-top boxes (STBs) coupled with TV screens, multimedia PCs, and network computers (NCs) are expected to be the appropriate CPE. The point of connection between the home network and the access network will be the digital RG. The RG may integrate the network terminator (NT) and will be capable of interfacing with multiple access and indoor media and terminals. Various approaches concerning the RG architecture and functionality may be considered.

A critical part in the networking architecture is the indoor network, as its cost is passed directly to the end user. Indoor alternatives may be organized in emerging or mature solutions that require new cabling and solutions with "no-new-wires" requirements. Ethernet is one of the major alternatives for the home network, as it is a mature technology, with proven speeds of up to 100 Mbps and simple, well-known installation and configuration procedures. Other alternatives include the IEEE 1394 protocol and technologies based on the existing home phone lines (HomePNA) or the power distribution network (powerline). Finally, radio-frequency (RF) technologies are considered among the best candidates for the home network [4].

In the following chapters, we will study these technologies and elaborate on their advantages and limitations.

## References

[1]   Zahariadis, T., et al., "Interactive Multimedia Services to Residential Users," *IEEE Communications Magazine*, Vol. 35, No. 6, June 1997, pp. 61–68.

[2]   Middleton, C., "Who Needs a 'Killer Application'? Two Perspectives on Content in Residential Broadband Networks," *Eleventh Australian Conference on Information Systems (ACIS2000)*, Brisbane, Australia, December 2000, at http://www2.fit.qut.edu.au/ACIS2000/ACIS%20papers/paper%20middleton.pdf.

[3]   Gupta, S., "Home Networking," white paper, Wipro Technologies, 2001, at http://www.wipro.com/shortcuts/downloads.htm.

[4]   Zahariadis, T., Pramataris, K., and Zervos, N., "A Comparison of Competing Broadband In-Home Technologies," *IEE Electronics and Communications Engineering Journal (ECEJ)*, August 2002, pp. 133–142.

# CHAPTER 3

# Wireline Home Access Network Alternatives

## 3.1 Introduction

For years the major barrier to the digital networked house has been the access network. The inadequate network infrastructure and the huge cost of new installations gave rise to the well-known last-mile problem, which hindered broadband access at home. Connecting each house to broadband access networks, however, represents an unprecedented opportunity to set the stage for a vast range of new home applications and value-added services for residential users and to expand the customer base beyond the corporate environment.

In the future, it is likely that a large number of customers will be connected to the information superhighway via fiber optics. However, at least for the next decade, several economic and geographical factos will prevent the introduction of dedicated fiber optics to the majority of customers. Instead, the evolution of existing access network technologies, such as asymmetric digital subscriber line (ADSL), HFC, wireless local loop (WLL), and passive optical networks (PON), is expected to be employed in the Physical Layer [1]. Most of these technologies have already been tested and evaluated, and their large deployment is already underway.

This chapter will identify and describe the main wired (or wireline) narrowband and broadband access network alternatives, ranging from analog modems that provide 28.8 Kbps to fiber-to-the-home (FTTH) solutions that provide multimegabyte solutions to home access requirements.

## 3.2 Analog Public Switched Telephone Network

For years, the sole available access solution to the home network has been the legacy public switched telephone network (PSTN). The PSTN is circuit switched, primarily designed for voice transmission.

Data traffic can be transmitted over PSTN by using analog modems. This is the simplest and most widely accepted narrowband solution for accessing the Internet, intranets, and remote local area networks (LANs) worldwide. It just uses the PSTN network and a pair of modems, one on each side. In most cases, for simplicity and space efficiency, at the Internet service providers (ISPs) side, instead of analog modems, access servers are used, which are able to concentrate and serve hundreds or thousands of connections. Since 1998, the analog modems have been able to provide up to 56 Kbps (28.8 to 56 Kbps), complying with the ITU V.90 standard. TheV.90 standard technology is asymmetric. The 56-Kbps data rate is only achieved downstream on a digital line from the ISP to the subscriber. The upstream connection is analog and provides data rates between 28.8 and 33.3 Kbps. Since early 2000, the ITU V.92 standard has provided a symmetric 56-Kbps transfer rate.

To realize 56-Kbps throughput, V.90/V.92 modems or access concentrators that use compatible modulation techniques at each end of the connection have to be installed. In cases where a modem with less bandwidth capability has been installed on one side, the communication will drop down to the highest common speed that both sides can synchronize to. However, even if V.90- or V.92-compliant modems are installed, the modems will achieve top speeds only if the connection has just a single analog to digital conversion, while the actual throughput is finally determined by the quality of the network line [2].

## 3.3   Integrated Services Digital Network

Before the Integrated Services Digital Network (ISDN), dedicated networks were required to provide services of a different nature and with different transmission characteristics, such as plain old telephone service (POTS) analog service, packet service, telex, data service, and the like. Dedicated and isolated network requirements have led to a number of drawbacks including high cost, low efficiency, and inconvenience. ISDN is an alternative technology that provides integrated voice and digital data services over regular phone lines.

ISDN supports two types of communication paths:

- *B-Channel*: The B-channel (or Bearer Channel) is a 64-Kbps channel, which transfers voice, data, video, or multimedia traffic. More than one B-channel can be aggregated for higher bandwidth applications.
- *D-Channel*: The D-channel (or Delta Channel) is a 16-Kbps or 64-Kbps channel, which transfers signaling and control data between ISDN switches and ISDN terminal equipment.

According to application bandwidth requirements, ISDN service is mainly available in two configurations:

1. *Basic-Rate Interface (BRI)*: BRI is the most widely available ISDN service for small business and residential customers. BRI is often called "2B+D" configuration, because it carries 2 B-channels ($2 \times 64$ Kbps) and 1 D-channel (16 Kbps). Thus, the total bandwidth of BRI is 144 Kbps, but only 128 Kbps are available for data and voice communications. A single BRI line can support up to two voice, fax, or data parallel sessions, and one packet-switched data session, which is used for the ISDN signaling. Multiple channels or even multiple BRI lines can be combined into a single faster connection (e.g., in a multimedia application) or split into individual channels for applications like normal voice or data transmissions.

2. *Primary-Rate Interface (PRI)*: PRI is a broadband configuration used by corporate enterprises. In Europe, Central/South America, and Asia, PRI is carried over E1 lines and contains 32 channels: 30 B-channels, 1 64-Kbps D-channel (30B+D), and 1 64-Kbps control channel, which is used mainly for framing, synchronization, and alarm transportation. Thus, the total E1 bandwidth is up to 2 Mbps. In the United States, PRI connection is carried over DS1 (or T1) lines and provides 23 B-channels and 1 D-channel (23B+D), so approximately 1.5 Mbps.

Figure 3.1 shows an example configuration for an ISDN basic-rate access network. The ISDN line terminates from the exchange office side at a device called the *line terminator* (LT) and from the customer premises side at a device called *network termination 1* (NT1). NT1 performs Physical Layer functions such as signal synchronization and conversion from a two-wire reference point called the *U Interface* to a four-wire reference point called the *S/T Interface*. An intelligent device at the customer premises, called *network termination 2* (NT2), performs Data Link Layer and Network Layer functions and provides a T Interface. A device that performs the combined functions of NT1 and NT2 is called *the network termination* (NT). Many ISDN terminal adapters, ISDN routers, or gateways integrate NT1, simplifying the

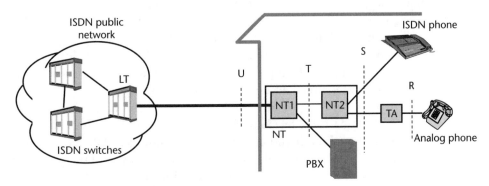

**Figure 3.1**  ISDN basic-rate network architecture.

installation and reducing the overall ISDN setup cost. However, a separate NT1 is more flexible in that it can support multiple ISDN devices, such as an ISDN Private Branch Exchange (PBX) with multiple ISDN phones.

According to ISDN specification, terminals are grouped as ISDN equipment (e.g., ISDN phone, ISDN PC adapter), called *TE1*, and non-ISDN devices (e.g., an analog phone), called *TE2*. TE1 devices can be connected directly to the NT, while TE2 devices are connected via a terminal adapter (TA) that provides for adaptation and ISDN protocol translation at the R interface.

In the case of home networking, ISDN has offered a relative cheap, reliable, and flexible solution. Moreover, even BRI ISDN offers the residential user the ability to access the Internet or Intranet from a single telephone line without interruption to voice service. Unlike a standard phone line, most ISDN providers do not provide the system power on the ISDN line. Thus, an analog phone line may be available for use as backup or lifeline service during emergency power outages [3].

## 3.4  xDSL

Normally, analog modems operate at voice frequencies and follow the Shannon-Hartley rule. The rule defines that the capacity of the channel $C$ is given by the following equation:

$$C = B \cdot \log_2 \left(1 + S/N\right)$$

where $B$ is the channel bandwidth, $S$ is the signal strength, and $N$ is the noise. For example, a channel with a bandwidth of 3,000 Hz and a signal-to-noise (S/N) ratio of 30 dB (1,000:1) will have a theoretical capacity of 30 Kbps. In real conditions, this modem would not provide more than 28.8 Kbps [4].

Digital subscriber line (DSL) technologies are the evolution of the telephone network. DSL services are dedicated, point-to-point (PtP), public network access technologies that, according to the type, technology, and operational environment, allow existing unshielded twisted-pair (UTP) copper local loops ("last mile") between the network provider's central office and the subscriber's premises to carry higher-bandwidth digital signals and multiple forms of data, voice, and video services in bit rates of up to 52 Mbps. Moreover, DSL is always connected and the user is able to have uninterrupted access to Internet/intranet.

The first type of DSL was invented by Bellcore Laboratories in the United States in the late 1980s [5] as the telecom operators' answer to the cable network providers in the demanding broadband market competition. Via DSL, telecom operators would have been able to provide, in a cost-effective way, VoD services over copper wiring. However, the real DSL boost came with the Telecommunications Reform Act of 1996, which allowed local-phone and long-distance carriers, cable operators, radio and television broadcasters, ISPs, and telecom equipment manufacturers in the United States to compete in one another's markets. Almost 20 years after the first DSL demonstration, VoD services have not yet been fully deployed and may never generate the initially foreseen revenues. However, many variations of DSL, referred to as xDSL, have been invented in the race to broaden cost-effectively home and residential connectivity to the Internet, which is the new killer application.

### 3.4.1  Transmission Challenges

By using higher frequencies over plain old twisted pairs, the xDSL technologies have to face various transmission challenges. The impairments, attenuation, and loss of signal depend on many issues including the physical characteristics of the wires, the distance from the central office, and the interference from the operational environment [6]. The main issues that affect xDSL transmission include the following:

- *Propagation Loss:* Propagation loss depends on the transmission frequency, the distance form the central office, and the physical characteristics of the copper line. For a frequency $f$ above 500 Hz, the propagation loss $Lp(f)$ over a distance $d$ is expressed in decibels by the following equation:

$$Lp(f) \approx a \cdot d \sqrt{f} + d \cdot b \cdot f$$

where $a$ is a function of the gauge of the loop and $b$ depends on the dielectric used for insulation [7].
- *Far-End Crosstalk (FEXT):* This type of interference occurs when two or more signals transmitted in the same direction on different UTP pairs have overlapping spectra. For example in Figure 3.2, transmission from UTP pair $i$ affects transmission from UTP pair $j$. The signal $D_j'$ is the attenuated signal $D_j$ affected by the interference signal $D_{i\text{-FEXT}}$.
- *Near-End Crosstalk (NEXT):* This type of interference occurs when there is a spectral overlap between the spectra of signals traveling in opposite directions on different pairs in a UTP cable. NEXT is far more damaging than FEXT and should be avoided if possible.

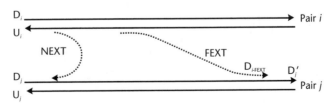

**Figure 3.2**  UTP crosstalk impairments.

- *Mismatched Impedances:* Mismatched impedances can occur at all connecting points in the communication link. The most damaging mismatches, however, tend to occur at the connecting point between the twisted pair and the drop wire, with almost double characteristic impedance.

- *Noise:* Noise can be categorized as radio frequency interference (RFI) and impulse noise. RFI is generated by signals emanating from broadcast systems (e.g., AM, short-wave radio), while telephone signals and indoor light dimmers mainly generate impulse noise.

For all of these reasons, the exact bandwidth provided by xDSL technologies is not always guaranteed, but depends on the specific network topology. To overcome these transmission challenges, two major modulation schemes, the single-carrier and the multicarrier modulation have been implemented. The single-carrier approach handles the complete frequency spectrum as a single channel. Frequency bands that carry traffic from the central office towards the subscriber (downstream) overlap with bands that carry traffic in the reverse direction (upstream), and they are separated by means of local echo cancellation, the same technique used by V.32 and V.34 modems. Thus, the single-carrier approach simplifies channel filtering, but requires more complicated processing to overcome the noise and interference. The multicarrier approach defines multiple carriers and channels using frequency division multiplexing (FDM) and handles each one independently. In many xDSL types, the downstream path is additionally divided by time division multiplexing (TDM) into one or more high-speed channels and one or more low-speed channels. In this way, the multicarrier approach requires more complex filters, but achieves more robust and higher-speed implementations.

Figure 3.3 shows the effects of the channel in a multicarrier signal. The propagation and FEXT attenuate the signal as a function of the frequency. However, mismatched impedance and noise may drastically suppress the signal.

As compared with the bandwidth and distance capabilities, there are many types of xDSL, and each has various subcategories according to the adopted technology. In the next sections we are going to describe the major xDSL types.

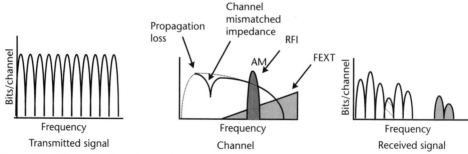

**Figure 3.3** Channel effects in a multicarrier signal.

### 3.4.2 ADSL

ADSL is the xDSL technology that is receiving the most attention, as it can provide VoD-like services over plain twisted-pair copper lines; thus, it can directly compete with cable modems. Statistically, home applications like VoD, Internet surfing, Intranet, remote LAN and e-mail access generate more traffic in the downstream direction than in the upstream. Home users normally receive, or download, much more data than they send. ADSL is based on this assumption and provides asymmetric bandwidth capabilities. According to the ADSL technology, speeds of up to 8 Mbps (even 12 Mbps in some cases) downstream and up to 1.5 Mbps upstream can be achieved at distances of 10,000 ft (3.3 km) or up to 1.5 Mbps downstream and 640 Kbps upstream at distances of up to 18,000 ft (5.5 km) using standard 24-gauge wire.

The ADSL architecture is shown in Figure 3.4. At the central office, a DSL access multiplexer (DSLAM) terminates the ADSL lines and splits the voice from the data traffic. POTS is forwarded to the public telephony network, while the data traffic is forwarded to the wide area network (WAN) backbone network. At the customer premises, a POTS splitter separates the regular phone signal from the data traffic, and an ADSL modem concentrates the in-home LAN traffic. For simplicity and reduced cost, the splitter and the ADSL modem may be integrated into an ADSL RG.

To transmit data without interrupting POTS services, ADSL uses higher frequencies. As Figure 3.5(a) shows, POTS occupies the frequencies of up to 3.4 KHz, while ADSL occupies the frequencies from 25 KHz to 1.1 MHz. It has been also proposed to avoid overlapping between ADSL and ISDN. In that case, as Figure 3.5(b) shows, ADSL occupies frequencies over 75 KHz, which are located by ISDN.

Two competing modulation schemes have been proposed for ADSL transmission: carrierless amplitude phase (CAP) modulation, which corresponds to the single-carrier approach, and the discrete multitone (DMT) modulation, which is the multicarrier approach. Both CAP and DMT use quadrature amplitude modulation (QAM) to separate digital carrier signals that occupy the same transmission bandwidth, but DMT divides the available frequencies into 256 subchannels, or tones,

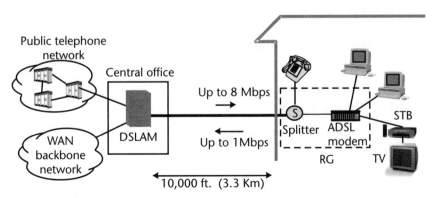

**Figure 3.4** ADSL network architecture.

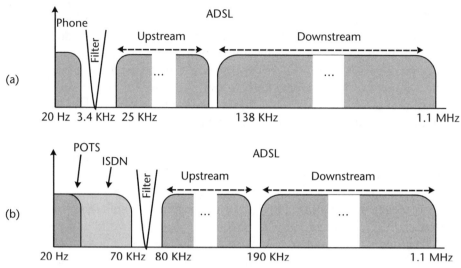

**Figure 3.5** ADSL spectrum allocation over (a) POTS and (b) ISDN.

and handles each one independently. CAP was deployed before DMT and has captured a large percentage of the xDSL market. However, it does not achieve interoperability between products from different manufacturers. On the other hand, DMT was adopted by the International Telecommunications Union (ITU) as the G.992.1 or G.dmt specification in June 1999 and later by the American National Standards Institute (ANSI) and the European Telecommunications Standards Institute (ETSI). Yet, the modulation schemes vary between vendors, and interoperability testing is required between DMT-based products.

ADSL installation is still quite expensive, as it requires a technician to install the POTS splitter. Splitterless G.dmt is a new nonstandardized effort to provide full-rate ADSL without the need for a splitter. As Figure 3.6 shows, a simple microfilter is installed between the legacy voice devices (e.g., analog phones or faxes), while a simple ADSL modem may interface with the network. The drawback of this solution is that the bandwidth capabilities are lower than with the splittered G.dmt, while just like with ISDN, an emergency lifeline is required in case of power outage.

### 3.4.3   xDSL Variations

Apart from the ADSL, there is a large list of xDSL variations, which are summarized in this paragraph:

- *ADSL Lite* or UADSL or G.lite is a lower-speed ADSL variation capable of providing rates of up to 1.5 Mbps downstream and 512 Kbps upstream. G.lite

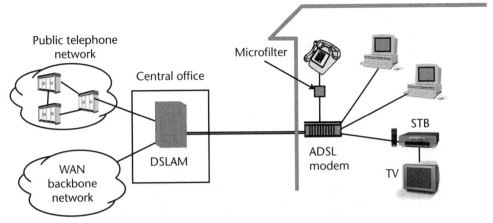

**Figure 3.6** Splitterless ADSL network architecture.

was designed to provide a path for evolution to full-rate ADSL. By reducing the data rate, line interference is manageable and, therefore, POTS splitters are not required. As compared with the G.dmt, G.lite provides easier installation. As there is no need for a POTS splitter, it involves cheaper ADSL modems, and it requires simpler processing and focuses broader markets since it can operate over longer distances than full-rate ADSL. G.lite was standardized as ITU standard G.992.2 in October 1998.

• *Rate-Adaptive Digital Subscriber Line (R-ADSL)* is an adaptive ADSL variation. The maximum transmission rate that R-ADSL is able to provide is the ADSL rate. However, the transmission rate may be adjusted dynamically. Adaptation may be the result of a signal from the central office or based on the specific twisted-pair access-line characteristics.

• *ISDN Digital Subscriber Line (IDSL)* is a symmetric service at ISDN speed. It provides full-duplex data-only service throughput at speeds of up to 144 Kbps. IDSL uses the ISDN modulation (2B1Q), but as with DSL service, IDSL is always connected; thus, it can use all available 144 Kbps. IDSL operates at distances of up to 18,000 ft, but may use standard U loop ISDN repeaters to extend the range to 36,000 ft.

• *High Bit-Rate Digital Subscriber Line (HDSL)* is a symmetric service, providing 1.544 Mbps over two copper pairs and 2.048 Mbps over three pairs. Due to its data rates, HDSL is often used as an alternative to T1 lines (in North America) and E1 lines. HDSL operates at distances of up to 15,000 ft (~4.5 km), but its range can be extended with low-cost repeaters. The major disadvantage of HDSL is that it requires multiple copper pairs. HDSL II has been proposed as the next generation of HDSL and will offer the same performance over a single pair.

• *Single-Line Digital Subscriber Line (SDSL):* Just like HDSL, SDSL is a symmetrical service at TI/E1 rates, but it uses a single copper-pair wire at distances of up to 22,000 ft (6.7 km). With SDSL, customers can initially purchase 128-Kbps capacity, and as their bandwidth needs grow, they can increase the line speed up to 1,544 Mbps without additional hardware investments. HDSL II is going to be based on SDSL.

### 3.4.4   Very High Bit-Rate Digital Subscriber Line

Very high bit-rate digital subscriber line (VDSL) is an xDSL technology that requires special attention. VDSL is the fastest xDSL technology, able to support up to 13-Mbps symmetric traffic or up to 52-Mbps downstream and up to 2.3-Mbps upstream traffic over a single copper-pair wire at a range of 1,000 ft (~0.3 km) to 4,500 ft (~1.3 km).

The large bandwidth capabilities enable VDSL to offer services like HDTV, multiple VoD, switched digital video, and broadband LAN services. The short operating distance differentiates VDSL from other xDSL technologies, and turns VDSL to an alternative to FTTH technologies. Figure 3.7 shows two VDSL network configurations. In both cases, signals are transmitted over fiber optics from the central office to an optical hub close to the subscribers. At the optical hub, a VDSL modem transfers the signal to the customer. In Figure 3.7(a), a passive NT has been selected that distributes all incoming signals inside the building in a tree- or bus-like topology. VDSL modems in each apartment connect the terminals or the indoor LAN to

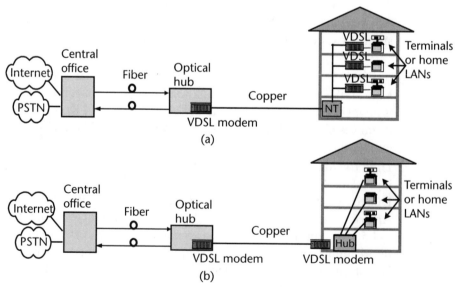

**Figure 3.7**   VDSL network topologies: (a) passive NT, and (b) active NT.

the VDSL network. In Figure 3.7(b), the VDSL is terminated at the ground floor and then an active hub concentrates the in-building traffic.

To achieve the high data rates, VDSL uses frequencies higher than ADSL that go well beyond 1 MHz. VDSL has not been standardized yet; thus, there are many different approaches to the spectrum allocation and the modulation type. Figure 3.8 shows a potential VDSL spectrum allocation scheme in which VDSL does not overlap with POTS and ISDN frequencies; thus, POTS or ISDN services may be used without interruption.

The greater problem with VDSL technology, especially with the downstream direction in case of symmetric services, is that frequencies over 1 MHz introduce very large impairments and high impulse noise. Therefore VDSL modems require large processing power and operate over short distances. To solve this problem, one of the proposed solutions is a DMT variation called *Zipper*. Zipper divides the frequencies into multiple narrow channels and assigns the channels to downstream and upstream directions in pairs. As Figure 3.9 shows, Zipper pairs guarantee that adequate bandwidth will be available in both directions.

## 3.5  Hybrid Fiber Coaxial

Another wireline home access alternative in many countries, especially in Northern Europe and in the United States is the CATV network, which was initially designed to broadcast only analog TV signals from the service-provider headend to subscribers. The incorporation of fiber lines has evolved the CATV to broadband HFC plants, while enhancements in modulation and use of unused spectrum have enabled the transmission of telephony and bidirectional broadband digital signals over the legacy CATV network.

A typical HFC network architecture is shown in Figure 3.10. The analog CATV signal in the central office or headend is multiplexed with bidirectional voice (telephony), Internet traffic, or digital video [e.g., VoD or digital video broadcasting

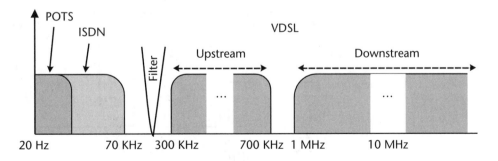

**Figure 3.8**  Potential VDSL spectrum allocation.

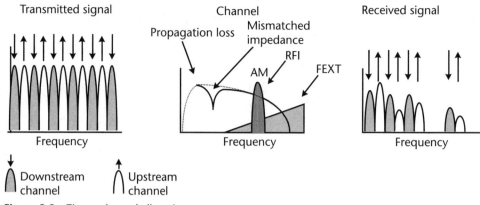

**Figure 3.9**   Zipper channel allocation.

(DVB)]. The cable modem termination system (CMTS) interconnects the HFC network with the PSTN/ISDN network, the Internet, and various service providers' networks.

The multiplexed signal is transmitted over the fiber lines to an optical hub, where it is converted from an optical to an RF signal. A typical optical hub is able to support up to 2,000 subscribers. The coaxial access network plant has a tree-like topology. The root is the optical hub, and the nodes are coaxial taps. Each tap concentrates multiple coaxial segments, and each segment connects multiple houses in a bus-like architecture. At the residential subscriber side, a cable modem (CM) splits the RF signal to analog CATV and bidirectional telephony or digital traffic. For a detailed description of the in-home coaxial network, please refer to Chapter 5.

Several organizations, including Digital Video Broadcasters (DVB), the Digital Audio Video Council (DAVIC), and the IEEE 802.14 Group, have tried to

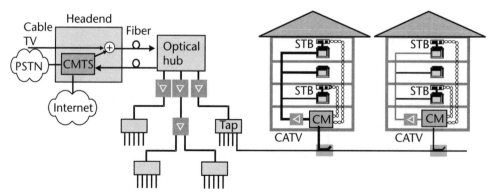

**Figure 3.10**   HFC/CATV network architecture.

standardize digital data services over HFC networks. However, delays in standardization led cable operators in the United States to adopt a de facto industry standard, the Data over Cable Service Interface Specification (DOCSIS), in 1995. Soon a European version of the standard, EuroDOCSIS, was announced. Today more than 50% of the cable Internet subscribers worldwide are using DOCSIS-based cable modems.

### 3.5.1  DOCSIS

DOCSIS was initially developed by Cable Television Laboratories, and it is the most widely adopted standard for bringing multiple Transmission Control Protocol (TCP)/IP sessions and Internet access to subscribers over CATV networks.

DOCSIS is an RF-based system designed to overlay CATV networks [8]. It is capable of transferring full-duplex traffic of up to 30 Mbps bidirectionally, while the broadcasting downstream capability can be up to 100 Mbps. However, service providers normally offer up to 6-Mbps downstream and up to 1-Mbps upstream traffic to allow for overallocation and return on investment.

#### 3.5.1.1  Physical Layer

The DOCSIS Physical Layer specification covers transmission for both the downstream (from the headend to the subscriber) and the upstream (from the subscriber to the headend) directions. The downstream traffic is transmitted at higher frequencies than the frequencies assigned to the analog CATV channels (from 50 MHz to 1 GHz), and the upstream traffic is transmitted below the TV channels (from 5 to 45 MHz). DOCSIS spectrum allocation is shown in Figure 3.11.

The downstream traffic is divided into channels of 6-MHz bandwidth and is QAM modulated with 64 (64 QAM) or 256 (256 QAM) constellation points. Many users (according to the service provider's policy, up to 500) share the same channel. The upstream traffic is divided into channels of 0.2-, 0.4-, 0.8-, 1.6-, or 3.2-MHz bandwidth and is QPSK or 16 QAM–modulated.

DOCSIS contains several features to assist with the mitigation of noise and interference [7]. Both downstream and upstream signals provide for forward error correction (FEC) to enable low bit-error rate (BER), while the DOCSIS hub controls the frequency, data rate, timing, equalization, and output power of each subscriber.

#### 3.5.1.2  Medium Access Control

The DOCSIS Medium Access Control Protocol (MAC) offers point-to-multipoint (PtMP) communications employing a continuous broadcasting signal in the

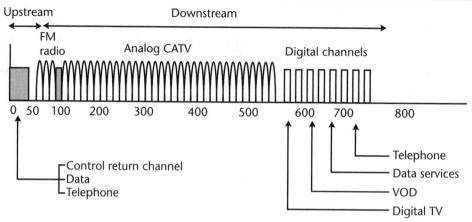

**Figure 3.11**  DOCSIS spectrum allocation.

downstream direction and a burst signal in the upstream direction. DOCSIS-compliant modems are capable of automatically scanning a range of downstream frequency channels, selecting the proper signal, synchronizing with it, and selecting the data that targets the specific modem.

During initialization, the modem obtains upstream parameters, automatic adjustments, transfer operational parameters, and the time of day and establishes IP connectivity. Multiple access in the upstream direction is provided by a combination of frequency division multiple access (FDMA) and time division multiple access (TDMA) mechanisms. Under this scheme, the CMTS assigns to each CM one channel and allocates, upon request, dedicated time slots to the CM on that channel, to enable the transmission of data without collisions. The CMTS also provides the time reference and controls the allowed usage for each time slot.

When the CM has data to transmit, it requests a granted period from the CMTS. The request may be either piggybacked on an upstream message or transmitted within a congestion period. During the contention period, all CMs are able to transmit and collisions are resolved using a combination of a binary exponential back-off period and an acknowledge-based mechanism. When the CMTS receives a request for a granted period, it assigns dedicated time slots to this CM.

DOCSIS 1.1 and 2.0 enable minimum data rates to individual subscribers and QoS guarantees greatly increased scalability and improved customer satisfaction with service quality.

### 3.5.1.3  DOCSIS in Wireless Environments

Apart from HFC networks, DOCSIS-based protocols and mechanisms are also used in wireless environments. As in cable networks, the DOCSIS modem and the CMTS cooperate with the radio system to locate the upstream and downstream channels dynamically in the intended frequency band. Additionally, if the wireless antennas are sectored, frequencies can be reused, increasing the effective bandwidth and the subscriber capability. The first wireless application for DOCSIS was in the multichannel multipoint distribution system (MMDS), or "wireless cable" systems (see Chapter 4), but satellite distribution soon captured the TV distribution market.

## 3.6 FTTx

Due to the durability and low-maintenance characteristics of fiber outside the plant and to fiber's wide bandwidth capability, network service providers prefer to install fiber as close to the customer as cost allows.

In general fiber optics are organized in a hierarchy like the one shown in Figure 3.12. Wide area fiber-optic networks are based on PtP connections between cities and offer up to 3–10-TB bandwidth [9]. Metropolitan core networks and access networks are in most cases ring topologies, interconnected via optical add-drop muliplexers (ADMs). In the lower layer, the access and consumer-premises networks may contain any combination of optical rings, optical links to buildings or neighborhoods, or HFC or hybrid fiber-copper links.

Focusing on the residential consumer market, the simplest solution for broadband service provision over fiber is the passive optical network (PON). The general architecture of a PON is shown in Figure 3.13. Fiber is extended between an optical line termination (OLT) at the central office or at a local exchange and an optical networking unit (ONU) on the residential side. The OLT interfaces with the backbone network that provides the services to the users. Between the OLT and the ONU, optical splitters provide a passive PtMP topology and enable the connection of a number of ONUs to an OLT in a tree topology. The ONU can be combined with an NT, providing an optical network termination (ONT). The PON technology uses TDM. Recent advances in fiber optics have provided a new technology called *wavelength division multiplexing* (WDM), which has been applied in long-distance networks as Dense WDM (DWDM) and UltraDense WDM (UWDM), and in metropolitan area networks in the form of Coarse WDM. However, frequency splitters could eventually provide a WDM-based solution for the first mile.

There are a number of variations based on the point to which fiber extends. Some of them are the fiber-to-the-cabinet or -curb (FTTC), fiber-to-the-building (FTTB), and FTTH. Like the general term xDSL for DSL technologies, the general term encompassing these variations is FTTx.

- FTTC extends the fiber-optic infrastructure to the street cabinets of a neighborhood. In the cabinet there is an optical hub (or ONU) that transforms the optical to electrical signals and distributes them to the buildings using copper

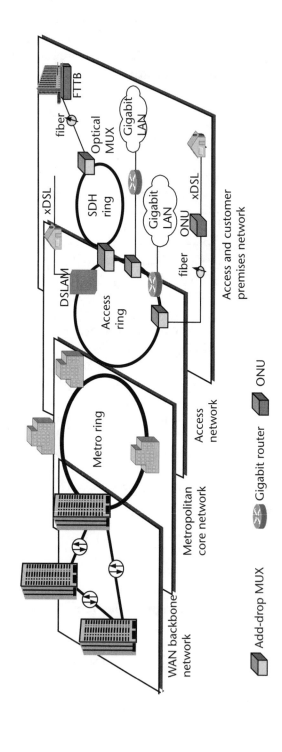

**Figure 3.12** Fiber optics network hierarchy.

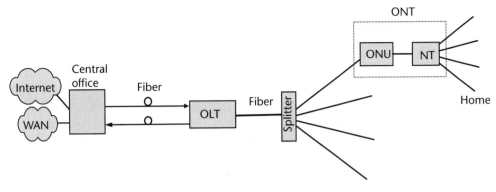

**Figure 3.13**  PON.

wiring. The last-meters selection may be Ethernet, providing capacity up to 100 Mbps, or VDSL providing capacity up to approximately 50 Mbps. In 2002, however, there were product announcements for a 100-Mbps Ethernet over 1 km of underground telephone cable.

- FTTH extends the fiber-optic infrastructure to subscribers premises, including SOHO users. FTTH offers the possibility of providing real broadband services with speeds of up to 2.5 Gbps per connection. Finally optical Ethernet technology is a low-cost alternative for FTTH provision. It provides up to 100 Mbps over fiber lines, and in the case of single-mode fiber, the distance from the central office can be up to 20 km.

- FTTB is a variation of FTTH, where traffic from multidwelling devices is concentrated and transmitted over fiber-optic lines.

FTTC, FTTB, or even FTTH may finally provide the solution to the last-mile bandwidth problem. However, the installation cost and the lack of a killer application that uses such a bandwidth requirement will discourage these solutions from the residential users' market for at least another 5 to 10 years.

## 3.7  Summary

In this chapter we presented the most prominent wireline access network technologies. Among them, analog modem and ISDN technologies are narrowband solutions, while xDSL, cable, and fiber solutions are broadband solutions.

Today, the most widespread technology is the analog PSTN, followed by the CATV, then ISDN, while some xDSL variations have already captured a significant market share. Especially since the differences between ISDN and xDSL are not many, ISDN is sometimes considered a DLS service. However, ISDN differs from xDSL in that it is a switched service, whereas xDSL is a PtP access service that provides uninterrupted connectivity. ISDN also requires external power for operation, while xDSL carries its own power on the line, and voice channels are not overlapped with data services. If a power failure occurs, xDSL data transmission is lost, but lifeline POTS still operates. On the other hand, DOCSIS is the leading platform for broadband last-mile delivery over a cable infrastructure. The scalability and flexibility of DOCSIS enable its deployment even over wireless systems, assuming the RF channel has a reasonably low BER.

Table 3.1 summarizes these wireline characteristics. Wireline technologies have already captured the largest market segment of broadband access networks. However, for reasons of economy and simplicity, new operators are expected to invest in wireless technologies. These technologies are analyzed in the Chapter 4.

**Table 3.1**  Summary of Wireline Access Technologies

| Access Technology | Bandwidth | Distance from the Central Office |
|---|---|---|
| V.90 analog modem | 56 Kbps downstream<br>Up to 33.6 Kbps upstream | No limits |
| V.92 analog modem | Symmetric, 56 Kbps | No limits |
| ISDN BRI | Symmetric, 128 Kbps | 18,000 ft (5.5 km) (can easily be extended) |
| ADSL | Up to 8 Mbps downstream<br>Up to 1.5 Mbps upstream | Up to 18,000 ft (5.5 km) (shorter for higher speeds) |
| R-ADSL | Rate-adaptive DSL based on channel conditions<br>Up to 1.5 Mbps downstream<br>Up to 512 Kbps upstream | Up to 18,000 ft (5.5 km) (shorter for higher speeds) |
| G.lite | Up to 1.5 Mbps downstream<br>Up to 512 Kbps upstream | Up to 18,000 ft (5.5 km) |
| IDSL | Symmetric, 144 Kbps | Up to 18,000 ft (6.7 km) (up to 36,000 ft with repeaters) |
| HDSL | Symmetric, E1-like (2,048 Mbps) with three copper pairs or T1-like (1.54 Mbps) with two copper pairs | Up to 15,000 ft (4.5 km) |
| SDSL | Symmetric, E1-like (2,048 Mbps) or T1-like (1.54 Mbps) | Up to 22,000 ft (6.7 km) |
| VDSL | Symmetric, 12 Mbps<br>Asymmetric, up to 52 Mbps downstream and 2.3 Mbps upstream | Up to 4,000 ft (1.2 km) |
| Cable network | Up to 30 Mbps downstream shared bandwidth<br>Up to 10 Mbps upstream | Up to 30 mi (can be extended to 200 mi) |
| FTTC | Up to 100 Mbps or 52 Mbps using Ethernet or VDSL in the last part | Does not cover up to 1 km from the house |
| FTTH/FTTB | Up to 2.5 Gbps<br>Up to 100 Mbps using Optical Ethernet | Up to 20 km |

# References

[1]   Zahariadis, T., et al., "Interactive Multimedia Services to Residential Users," *IEEE Communications Magazine,* Vol. 35, No. 6, June 1997, pp. 61–68.

[2]   Aber, R., "xDSL Local Loop Access Technology," technical paper, 3Com, March 1999, http://www.adimpleo.com/library/3com/500624.pdf.

[3]   Sunrise Telecom, "Technology Series: Introduction to ISDN—ETSI," publication number TEC-ISDN-0001 Rev. A, 2000, at http://www.sunrisetelecom.com/technotes ISDN_IntlTechNote_Final.pdf.

[4]   Hawley, G., "System Considerations for the Use of xDSL Technology for Data Access," *IEEE Communications Magazine*, March 1997, pp. 56–60.

[5]   Zervos, N. A., and Kalet, I., "Optimized Decision Feedback Equalization Versus Optimized Orthogonal Frequency-Division Muliplexing for High-Speed Data Transmission over the Local Cable Network," *ICC'89*, Boston, June 1989, pp. 1080–1085.

[6]   Lawrence, V., et al., "Broadband Access to the Home on Copper," *Bell Labs Technical Journal,* summer 1996, pp. 100–114.

[7]   Im, G., and Werner, J. J., "Bandwidth Efficient Digital Transmission over Unshielded Twisted-Pair Wiring," *IEEE Journal on Selected Areas in Communications,* Vol. 13, No. 9, Dec. 1995, pp. 1643–1655.

[8]   Cable Television Laboratories, "Data-Over-Cable System Interface Specifications. Radio Frequency Interface Specification," at http://www.cablemodem.com/Specs/SP-RFI-I06-010829.pdf.

[9]   DAVIC, "Delivery System Architecture and Interfaces," DAVIC 1.3, Part 4, Revision 6.2, Geneva, Switzerland, 1997.

# CHAPTER 4

# Wireless Home Access Network Alternatives

## 4.1 Introduction

Privatization and deregulation of the telephone and telecommunications industry has increased the competition among telephone operators, service providers, and utility companies over broadband access and integrated multiservice offerings. Broadband wireless technology appears to be one of the most viable solutions, as it enables infrastructure to be rolled out incrementally and for new services such as high-speed Internet access and video distribution to be turned out rapidly [1].

Deployment of cable and fiber systems is particularly difficult in certain areas where installing underground infrastructure is either not viable owing to existing buildings and infrastructure in place or impractical owing to the terrain or landscape. It is also very expensive, requiring a huge initial investment and extensive build-out of the infrastructure before commercial services can be launched.

On the other hand, WLL provides a simple, fast, flexible, and cost-effective mean to set up new communication links for new customers. Once the network planning and layout are complete and the basic infrastructure has been installed, services can be deployed immediately. WLL also limits or eliminates the problems associated with obtaining rights of way, as well as the high expense of laying wire, cable, or fiber, while providing similar access bandwidth and two-way capability.

In this chapter we present two fixed-wireless technologies. MMDS, which mainly targets TV broadcasting and covers longer distances, and local multipoint distribution service (LMDS), which provides two-way communications across shorter distances. Finally, satellite communications are briefly described.

## 4.2 MMDS

MMDS, also referred as wireless cable, is a fixed-wireless technology that has been primarily used for analog TV broadcasting. MMDS is a line-of-site technology, with a typical cell radius of approximately 25 to 35 mi (up to 50 km) depending on the terrain and the antenna placement. It uses the 2.500- to 2.686-GHz frequency band, covering 186 MHz of spectrum, and the 2.150- to 2.162-GHz frequency band, covering an additional 12 MHz of spectrum. With each video channel allocating 6 MHz of spectrum in analog format, MMDS is capable of transmitting up to 33 video channels. However, MMDS operators are looking to use digital compression techniques to increase the number of channels to around 200, making the systems more competitive not only with the wired cable systems, where MMDS has the advantage of simpler and less expensive installation, but also with the satellite-delivered systems.

MMDS has been used for TV broadcasting since 1983, when the U.S. Federal Communications Commission (FCC) allocated frequencies for licensed network providers. Since the Telecommunications Act of 1996, which broadened competition in all area of telecommunications in the United States and worldwide, MMDS systems have been used to provide an alternative to new service providers and subscribers where access has been limited to DSL or cable. In September 1998, the FCC announced new rules that allow two-way service via MMDS frequencies. As a result, MMDS is able to provide broadband access for multichannel TV, data, voice, and Internet service.

The network architecture of MMDS is shown in Figure 4.1. Like broadcast television, MMDS is transmitted from a tower, usually located on a mountain or tall building, to antennas affixed to residences or businesses throughout a market. The signal is forwarded to the transmission tower or antenna by fiber or other broadband medium (e.g., PtP wireless repeater, satellite connection). To enable bidirectional wireless communication, for example to enable VoD-like applications or Internet or intranet access, a terrestrial wired network is required. A PSTN/ISDN or xDSL modem may provide for an upstream link, which is used to control the transmission and select the content.

MMDS covers a large area with a radius of up to 50 km. However, this coverage is very difficult to achieve, especially in dense areas, as MMDS requires a direct line-

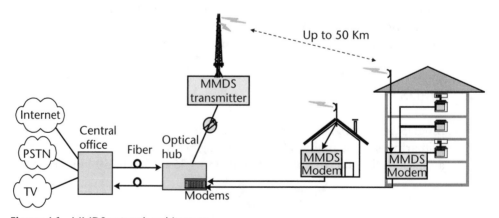

**Figure 4.1**   MMDS network architecture.

of-sight between the transmitting and receiving antennas, which results in the need for additional transmitters and repeaters. Another disadvantage of MMDS is cochannel interference from other cells. To reduce any chance of interference, FCC requires a minimum 35-mi protection zone between MMDS operators.

Still, MMDS is a cost-effective solution for digital TV broadcasting when compared with cable, xDSL or fiber, but satellites have already captured the largest percentage of this market.

## 4.3  LMDS

LMDS, also referred as the local multipoint communications system (LMCS), is a broadband, fixed-wireless access system, which allows for two-way digital communications for voice, broadcast video, VoD, and high-speed data communication, without the need for terrestrial wired networks to communicate back to the central office.

LMDS uses licensed frequencies in the range of 28 to 31 GHz. The selection of licensed spectrum and the large allocated bandwidth make the LMDS systems much less susceptible to interference and offer far greater speeds and communication links of 1 to 2 Gbps.

LMDS is a line-of-sight technology, which may be deployed in PtMP or PtP configurations. As Figure 4.2 shows, in case of PtMP communication, a central station (CS), which is connected to the backbone network, communicates with a number of terminal stations (TSs), which are located within the servicing range of the CS cell. If, in the line-of-sight propagation path between the CS and the TS, there is an obstruction that the signal cannot penetrate, such as a high building, group of trees, or hilltop, a reflector station (RS) has to be installed in an appropriate position to

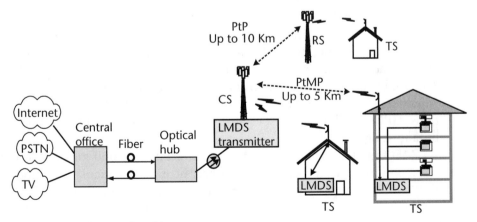

**Figure 4.2**   LMDS network architecture.

retransmit the signal and overcome the obstacle. PtMP connections are used for service distribution to customers and typically have a range of 3 mi (up to 5 km). PtP LMDS systems connect two locations and operate over longer distances, up to 10 mi (16 km). In practice, the PtMP systems are used for service provisioning, while PtP systems mainly constitute the LMDS backbone network.

LMDS is a very good alternative to wireline access networks. The short coverage range and the two-way communication capability make LMDS ideal for dense coverage areas. With proper planning and by implementing frequency reuse, the spectrum may essentially be split, allowing channels to be reused over and over in noncontiguous areas [2].

Of course, LMDS also has a number of limitations. Apart from line-of-sight, it requires careful planning to avoid the need for large numbers of repeaters. Moreover, LMDS signal strength is greatly reduced by the presence of moisture (rain fade). Channel fading due to heavy downpours for cells with a radius of less than 1 km is minor; however, these effects become more pronounced on longer path links, increasing the service outage probability due to a passing rainstorm. Regions have been characterized by rainfall patterns throughout the world, and certain locations are particularly devastated by this type of climatic condition [3]. Very heavy rainfall may cause the signal to be dropped completely [4]. Although these outages are uncommon and typically brief, they can make the service inappropriate for some critical applications.

## 4.4 Satellite

Home access via satellites has been another emerging solution. Communication satellites can be categorized into three groups: geostationary Earth orbit (GEO), medium Earth orbit (MEO), and low Earth orbit (LEO) [5]. The advantage of GEO satellites is that they rotate at the same angular velocity as the Earth, maintaining a fixed position with reference to the ground. In this way, GEO satellites appear at a fixed latitude and longitude. Moreover, due to their rather large distance from the Earth's surface (roughly 36,000 km), GEO satellites have a very large servicing area—almost one-third of the Earth's surface, from about 75°S to 75°N latitude. The combination of the fixed position along with the very large servicing area provides near-global coverage with a minimum of three satellites in orbit. Communications GEO satellites are especially useful for broadcasting services (e.g., TV broadcasting).

MEO satellites rotate at an altitude of around 10,000 km and do not have a fixed position over the Earth. They rotate at different angular velocities from the Earth; thus, they move with reference to the ground. A prime example of an MEO system is the U.S. Navistar global positioning system (GPS). Finally, LEO satellites rotate in orbits much closer to the Earth at a distance of 500 to 2,000 km above the surface. Like MEOs, LEO satellites do not have a fixed position over the Earth. The maximum time during which a satellite in LEO orbit is above the local horizon for an observer on the Earth is about 20 minutes, while there are long periods during which the satellite is out of view. This may be acceptable for a store-and-forward type of communication system, but not for interactive communications. LEO satellites, owing to their smaller distance from the surface, achieve very good end-to-end delay and have lower power consumption requirements for both the mobile terminal and the satellite. Thus, they are preferred for mobile communications [6]. For example, IRIDIUM is a LEO system that uses 66 satellites (plus 6 in-orbit spares) in six orbital planes and provides personal satellite communications.

In the case of broadband access via satellite, the GEO systems are preferred. Communication satellites typical lease channels to large corporations or large TV stations, which use them for long-distance telephone service, virtual private networks (VPNs), or TV distribution. Home access via GEO satellites follows a standard called *Direct Broadcast Satellite* (DBS). In DBS, the satellite operates as a microwave reflector that is able to deliver multimedia data to the home at speeds of 45 Mbps.

As Figure 4.3 shows, the signal is sent up to the satellite, where it is processed and sent back down to the Earth in a very broad beam of radio waves. At the customer's premises, a satellite dish is required to capture the signal. Normal satellite dishes have a diameter as small as 18 in. Moreover, one dish may support both single- and multiple-dwelling units. As the signal is broadcasted, encryption is applied to prevent authorized access to the content. Video broadcasting via satellite may provide the illusion of interactivity via switching between multiple broadcasting channels. However, real interactivity, for example Internet access, may be provided if the subscriber also has a terrestrial wired connection (e.g., a PSTN/ISDN modem to deliver the upstream/control channel).

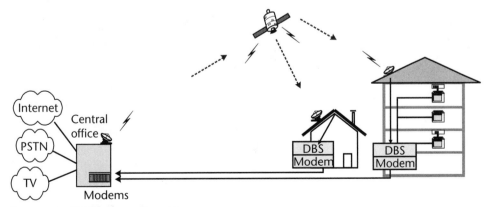

**Figure 4.3** Satellite network architecture.

The greatest advantages of satellite over any other medium are large coverage and large shared downstream bandwidth (up to 9 Gbps). However, the large initial cost, the lack of real interactivity, and the long round-trip delay limit satellite usage mainly to entertainment and video distribution applications.

## 4.5   Summary

In the future, it is likely that a large percentage of customers will be connected to the information superhighway via fiber optics. However, at least for the next decade, several geographical, economic, and political factors will prevent the introduction of dedicated fiber optics into the majority of customers' houses. Instead, several other technologies may be employed. In Chapters 3 and 4, some of the most widely accepted alternatives, such as xDSL, HFC, fixed wireless, optical networks, and satellite communications, have been presented. Table 4.1 summarizes each physical medium and its typical applications.

**Table 4.1**   Comparison of Access Network Technologies

| Access Type | Physical Medium | Typical Applications | Comment |
| --- | --- | --- | --- |
| POTS | Twisted-pair | Telephony, low-rate data | Uses standard telephone lines; has wide availability and low cost |
| ISDN | Twisted-pair | Telephony, medium-bandwidth data | Widely available; Telcos and ISPs have already invested and built out the infrastructure to develop it further |
| xDSL (HDSL, SDSL, ADSL, G.lite, VDSL) | Twisted-pair | Telephony, VoD, broadband data | Use existing twisted pair; G.Lite and SDSL might get the highest market share short term, ADSL and VDSL long term |
| HFC | Fiber, coax | Telephony, broadcast video, broadband data | Based on the existence of CATV Network; good especially for new builds and rebuilds |
| FTTH (FTTC) + VDSL | Fiber | Telephony, broadcast video, VoD, broadband applications | Long-term solution for broadband; cost is the obstacle, and deployment is still in the first phase |
| MMDS | Air 2–3 GHz | Video distribution | Good for rapid deployment of video overlays; offers PtMP applications; combines with other technologies (xDSL) for interactive applications |
| LMDS | Air 28–38 GHz | Broadcast video, telephony, broadband data | Good for rapid deployment of interactive communications; offers two-way PtP and PtMP applications |
| DBS | Satellite | Broadcast video, broadband data | Provides broad geographic coverage; support for local programming is difficult |

# References

[1] Dravopoulos, I., et al., "Adaptive Traffic Handling over Fixed Broadband Wireless Access Systems," *IST Mobile Communications Summit 2001 (MobileSummit 2001)*, Barcelona, Spain, Sept. 9–12, 2001.

[2] Zahariadis, T., Pramataris, K., Zervos, N., "A Comparison of Competing Broadband In-Home Technologies," *IEE Electronics and Communications Engineering Journal (ECEJ)*, Aug. 2002, pp. 133–142.

[3] Honcharenko, W., et al., "Broadband Wireless Access," *IEEE Communications Magazine*, Vol. 35, No. 1, Jan. 1997, pp. 20–26.

[4] Zysman, G., Thorkildsen, R., Lee, D., "Two-Way Wireless Broadband Access," *Bell Labs Technical Journal*, summer 1996, pp. 115–129.

[5] Zahariadis, T., et al., "Global Roaming in Next-Generation Networks," *IEEE Communications Magazine*, Vol. 40, No. 2, Feb. 2002, pp. 145–151.

[6] Werner, M., et al., "ATM Based Routing in LEO/MEO Satellite Networks with Intersatellite Links," *IEEE Journal on Selected Areas in Communications*, Vol. 15, No. 1, January 1997, pp. 69–82.

# CHAPTER 5

# Technologies That Reuse Existing Home Wiring

## 5.1 Introduction

In the previous chapters, we reviewed various wireline and wireless access network technologies that bring the information superhighway just outside of the majority of the houses in developed and developing countries worldwide. In the following chapters we will present the various in-home-networking technologies. In-home networks are organized into three groups [1]:

1. Network technologies that reuse the existing home wiring, namely power lines, phone lines, and coaxial cabling. The major advantages of these technologies are that they do not require rewiring of the buildings and they are directly applicable to new and old houses. The major disadvantages are their limited performance due to the networks' structure and the interference from the original operation of the network.

2. Technologies that require special network wiring, like special data cables or fiber (plastic or optical). These technologies are mature and offer enhanced performance, but their wiring requirements prevent their extended use in existing buildings.

3. Wireless technologies [e.g., radio frequency (RF) and infrared (IR)] that combine easy installation with enhanced performance, but in many cases entail line-of-sight requirements, limited coverage, or increased cost.

Structured wiring for power and telephone distribution represents the major in-home existing network, while a large percentage of new buildings also have coaxial cabling for in-home TV distribution. In this chapter, we present the major technologies that reuse these networks for data distribution.

## 5.2    Powerline Communications

Powerline communications is an emerging networking technology that reuses the house electrical wiring system to link appliances to each other and to the Internet. The initial idea behind powerline technology is to control and manage any device that is plugged into an outlet, including lights, sensors, coffee machines, alarm systems, and television sets. The next step is to use the electrical wiring system for high-speed data communications between intelligent electronic devices, like PCs or VCRs, which are already connected to ac outlets for gathering the electricity they need to operate. The market potential for such technology is huge, considering that the vast majority of houses worldwide is already wired with electricity lines.

The powerline network architecture is shown in Figure 5.1. The public distribution electricity network terminates in a power meter, which is also the demarcation point between the access and the in-home powerline network. The meter may also have circuit breakers to avoid voltage spikes or an RG may provide similar functionality. Consumer-electronic devices are connected to the in-home ac powerline network. Legacy PCs are connected to the powerline network for ac power supply. Moreover, a powerline modem or controller may transmit Ethernet or serial data, providing data networking and Internet access. Future PCs and modern consumer-electronic devices may integrate embedded circuitries into their power-supply modules, which would provide control and data communication directly over the powerline network.

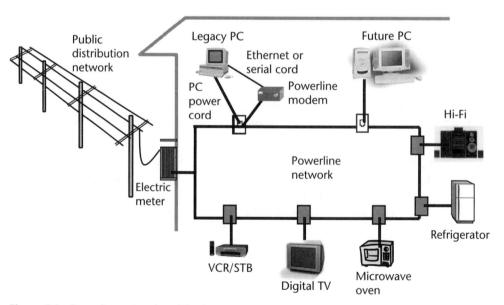

**Figure 5.1**    Powerline network architecture.

### 5.2.1 Obstacles Faced by Powerline Technologies

The major obstacles to transmitting data over the powerline network are the physical topology and the physical characteristics of electrical cabling and the interference generated by electric current transmission.

- *Signal Attenuation and Reflection Loss:* Attenuation describes how the signal strength decreases and loses energy as it moves across a physical medium. Wire nuts, switches, wall socket outlets, and multiple connected appliances cause impedance mismatches, which scatter the signal's power [2]. Unterminated points, changes in the physical wiring structure, and jumping phases present impedance discontinuities, which distract the signal and generate signal reflections that may even result in cancellation of the original data signal. Signal attenuation increases with the wire distance and with higher signal frequencies. Lower frequencies and isolated cabling reduce the attenuation and reflection losses respectively.

- *Network Interference:* Interference, one of the foremost obstacles to powerline communications, refers to any undesired signal that obstructs data signals. Undesired signals are called *noise*. There are two type of noise: impulse noise, which is generated by a precise source, and background or white noise, which is the result of interference from multiple sources. Impulse noise is mostly generated by devices that have ac motors (e.g., washing machines, vacuum cleaners, blenders, hair dryers), because when they are in operation, they inject impairments due to electrical spikes of noise. One of the methods for overcoming noise impairments is to increase the signal transmission frequency. However, this in return increases the signal attenuation. Thus, high-speed powerline technologies have to select an optimal transmission frequency that maximizes signal robustness, while keeping signal attenuation low.

### 5.2.2 X10

X10 is a rather low bandwidth protocol that provides up to 100 Kbps over the existing home powerline wiring tree by exchanging coded, binary, low-voltage signals superimposed over the 110V ac current. X10 technology was initially developed to integrate with low-cost lighting and appliance control devices. Since early 2000, advances in X10 have enabled higher speeds, targeting mainly broadcasting applications and communication between home PCs and controlled home appliances.

There are specialized X10-enabled appliances that integrate X10 transmitter modules and provide full remote control. However, X10 may offer simple control (e.g., on, off, dimmer) to any legacy electrical device connected to the in-house powerline network by using a specialized controller/transmitter device. The X10 controller is plugged into a standard electrical outlet, and the appliance is plugged into the controller. The controller communicates using the powerline network and forwards the control commands by varying the ac power to the legacy appliance.

X10 is a broadcast protocol that defines a communication language, which allows compatible home appliances to talk to each other based on assigned addresses. When a X10 transmitter sends a message, any X10 receiver plugged into the household powerline tree receives and processes it, and responds only if it carries its address. X10 enables carrying unique addresses up to 256 devices, while two or more devices can be addressed simultaneously if they are assigned the same address.

To differentiate the data symbols from current, X10 transmits short amplitude modulation (AM) bursts that represent digital information [3]. The receivers are synchronized at the zero-voltage crossing point of the 60-Hz ac sine wave and receive one bit on the cycle's positive or negative transition. To reduce errors, X10 uses two zero crossings to transmit a binary digit, and each packet contains two identical messages of 11 bits each. Each command usually contains two packets and a three-cycle gap between each packet. Thus, a complete X10 command requires 47 cycles.

As Figure 5.2 shows, each message starts with a code that is always the unique number "1110." Then a house code follows, which is up to four bits long. Finally, according to the message, a number code (1 to 16) or a function code (e.g., ON, OFF, Status Request) follows.

X10 basically targets control and automation applications, although entertainment products are already available. Moreover, wireless plug-in bridges may receive radio signals and connect the X10 signal to the power line, enabling the reception of X10 signals over the air.

### 5.2.3   Consumer-Electronic Bus

The consumer-electronic bus (CEBus) powerline carrier technology was one of the first attempts to transport messages between household devices, using the home's

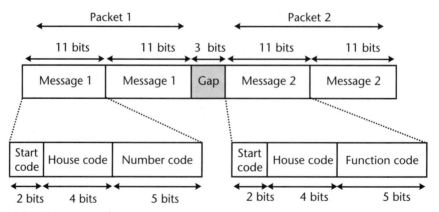

**Figure 5.2**   X10 command structure.

120V ac electrical wiring in a standardize way. The CEBus Protocol is an American National Standard (ANSI/EIA-600) [4] developed by the Consumer Electronics Association (CEA), a sector of the Electronics Industries Association (EIA). Back in 1984, the CEBus specification was released, defining an open architecture and protocols that enabled communication through powerline wires, low-voltage twisted pairs, coax, IR, RF, and fiber optics. More than 400 companies have attended the CEBus committee meetings, providing a comprehensive standard intended for the consumer-electronics industry.

The main objective of CEBus has been to standardize communications between consumer-electronics devices, while subobjectives have included the following:

- Low-cost implementation over a variety of physical media;
- Home automation for retrofit into existing cabling networks;
- Distributed communication and control strategy;
- Basic PnP functionality allowing devices to be added or removed from the network without interrupting the communication of other subsystems.

The CEBus/EIA-600 specification covered the following issues:

- The overall topology of the network and for each individual medium;
- The electrical and physical specifications and interfaces for each medium;
- The protocol for network access and the description of the control message format;
- A command language that allows all devices to communicate a common set of functions to be performed, called *common applications language* (CAL).

As Figure 5.3 shows, CEBus defines a full protocol stack with multiple physical interfaces and a vertical System Management Layer specialized for CEBus applications. Data packets are transmitted using spread-spectrum technology. At the Data Link Layer, the CEBus MAC protocol is based on a Carrier Sense Multiple Access/Collision Detection and Resolution (CSMA/CDCR) algorithm, similar to Ethernet. To avoid data collisions, CSMA/CDCR requires a network node to wait until it does not sense any traffic on the medium before sending a packet. The Presentation, Session, and Transport Layers are combined and provide very basic functionality.

In the Application Layer, CEBus has introduced CAL, which defines a common command syntax and vocabulary and enables devices to exchange commands and status requests. However, CEBus dealt only with home automation and speeds of up to 10 Kbps and never offered true multimedia capabilities. In late 1995, CEBus became part of an umbrella standard known as Home Plug 'n' Play (HPnP). The only part of CEBus that has not become outdated after merging with HPnP is CAL,

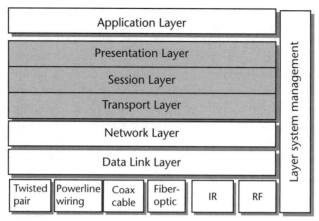

**Figure 5.3**   CEBus protocol stack.

which now promotes interoperability among various home-control technologies, such as X-10, IrDA, and audio-video (A/V) bus.

### 5.2.4   High-Speed Powerline Carrier

At the end of 1999, the CEA formed the Data Networking Subcommittee R7.3, and began work on a new high-speed powerline Carrier (PLC) standard. High-speed PLC technology aims to deliver burst data rates of up to 20 Mbps over powerline cables. However, just like CEBus, PLC shares the same power network with motors, switch-mode power supplies, fluorescent ballasts, and other impairments, which generate substantial impulse and wideband noise. Network loads create a time-varying environment, altering the impedance of the line and the noise environment. To face this mesh environment, the R7.3 subcommittee studied various technologies that take widely differing approaches depending on the applications they are pursuing [5]. Technologies and algorithms including orthogonal frequency division multiplexing (OFDM), rapid adaptive equalization, wideband signaling, FEC, segmentation and reassembly (SAR), and token-passing MAC Layer are employed to enhance transmission robustness, increase the required bandwidth, guarantee quality, and provide both asynchronous and isochronous transmission. However, high-speed PLC communications are still not mature enough for a large home-network market share.

### 5.2.5   HomePlug

The HomePlug Powerline Alliance is a nonprofit industry association established to provide a forum for the creation of an open specification for home powerline networking products and services [6]. HomePlug was formed in early 2000 to facilitate and promote the rapid availability, adoption, and implementation of cost-effective, interoperable, and specifications-based home powerline networks and products that would enable the connected home. Moreover, HomePlug aims to build a worldwide standard, pursuing frequency division for coexistence with access technologies in North America, Europe, and Asia.

The HomePlug Alliance consists of 13 founding members (3Com, AMD, Cisco Systems, Compaq, Conexant, Enikia, Intel, Intellon, Motorola, Panasonic, RadioShack, SONICblue, and Texas Instruments), while more than 80 member companies participate in the development of the specification and its promotion in the industry. The alliance aims to accelerate the demand for powerline products and services through the sponsorship of marketing and user-education programs.

HomePlug's final specification objective is a simple-to-use, Ethernet-class, powerline networking standard that will support a range of products for the gaming, consumer electronics, voice telephony, and PC markets. As a starting point, HomePlug evaluated various powerline technologies in an industrywide, open evaluation process that incorporated theoretical analysis and lab and field testing. The evaluation criteria included Ethernet-class speed, clear compatibility, robustness, and ease of implementation. Among the evaluated technologies, HomePlug selected Intellon's technology as the baseline upon which to build the HomePlug 1.0 specification. In early 2001, HomePlug began field-testing its baseline technology specification in more than 500 consumer households and on nearly 10,000 wiring paths in the United States, Canada, Japan, Korea, and several European countries. Upon completion of the field trials, HomePlug will finalize its initial technology specification and establish a certification lab to guarantee product compliance.

The HomePlug Alliance keeps proprietary the methods of scaling HomePlug technology to higher speeds. However, its areas of focus are modulation techniques, protocol enhancements, and circuit-design optimization.

Last but not least, the powerline network can also be used as an access network. However, the different characteristics, devices, and architectures of the network worldwide, along with the highly noisy environment, limit short-term expectations.

## 5.3   Phone-Line Communications

The second most widespread structured wiring network for the home is the telephone network. Using the phone line with a modem is also the most popular way to get access to the Internet. This section discussed phone-line communication technologies, which aim to use this network for in-home data communication.

Figure 5.4 shows the phone-line network architecture. In this figure, in-home devices (e.g., telephone, Web telephone, PC, security camera, VCR) are interconnected via the telephone lines using R-J11 sockets. Moreover, connecting R-J11 to

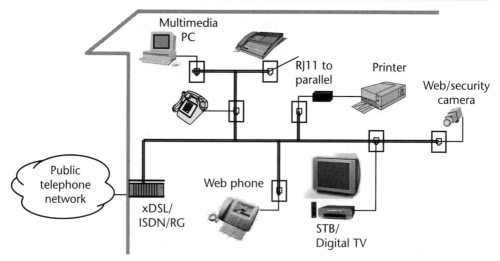

**Figure 5.4**　Phone-line network architecture.

parallel boxes may enable connection of devices that do not have a telephone inter-face (e.g., a printer).

The technological obstacles with the phone-line network are smaller as com-pared with the powerline network, because there are fewer devices connected to the phone-line network and they generate less signal attenuation, impedance, and inter-ference than electrical devices. On the other hand, the network topology is a random "tree," and sometime simple actions like unplugging a phone may change the tree structure. Another practical limitation is the number of telephone jacks and their location in the house. Houses in the United States tend to have multiple phone jacks, while households in other countries, particularly Europe, are often limited to one or two phone jacks. In addition, the location of those jacks with respect to the devices that need to be networked is another problem [7].

### 5.3.1　HomePNA

HomePNA is a de facto standard defined by the Home Phoneline Networking Asso-ciation [8] to promote and standardize technologies for home phone-line network-ing and ensure compatibility between home-networking products. The association was founded in 1998 by 11 companies [3Com, AMD, AT&T Wireless, Compaq, Conexant, Epigram, Hewlett-Packard (HP), IBM, Intel, Lucent Technologies, and Tut Systems], and by 2001 it had more than 130 member companies.

HomePNA takes advantage of existing home phone wiring and enables an immediate market for products with "Networking Inside." In late 1998, adapting the 802.3 framing and the Ethernet CSMA/CD MAC for phone-line networks,

HomePNA 1.0 was released. HomePNA 1.0 provided 1-Mbps data traffic mainly for control, home automation, and Internet-access applications over the home phone lines, without interrupting standard telephone service. In December 1999, HomePNA 2.0 was announced. HomePNA 2.0 provides up to 10-Mbps capacity, while future versions promise bandwidth of up to 100 Mbps.

HomePNA 2.0 specifies the two lower layers of the OSI protocol stack, namely the Physical and Data Link Layers. To enable data and voice services over a single piece of telephone wire, HomePNA uses FDM, which assigns each communications service a different frequency spectrum; thus, services do not overlap. As Figure 5.5 shows, telephone service is allocated the 20-Hz to 3.4-KHz range (in the United States; this range is slightly greater worldwide) and ADSL services are allocated the 25-KHz to 1.1-MHz freqencies. HomePNA allocates the frequencies between 4 MHz and 10 MHz [8].

HomePNA 1.0 used pulse position modulation (PPM) resulting in a 1-Mbps data rate. To provide 10 Mbps, HomePNA 2.0 uses QAM. Moreover, to achieve greater robustness, HomePNA 2.0 uses adaptive modulation. Instead of encoding a fixed number of bits per symbol, the transmitter may vary the packet encoding from two to eight bits per symbol on a per-packet basis. The packet's header is always encoded at two bits per symbol, so that every receiver can demodulate the packet's header and check the modulation of the rest of the packet. The 10-Mbps data rate is achieved with six-bits-per-symbol encoding.

The frame format of HomePNA 2.0 is shown in Figure 5.6. It is based on Ethernet format, but has a modified preamble and trail to accommodate enhanced robustness and adaptive modulation. The frame begins with a known, 64-symbol preamble. The preamble provides synchronization, robust carrier sensing, and collision detection. The Frame Control field is 16 symbols long. It contains an eight-bit frame type (only type = 0 is currently used, while the rest are reserved for future system frame formats), an eight-bit field that specifies the modulation format (i.e., bits per symbol), and other miscellaneous control fields, including an eight-bit cycle redundancy check (CRC) header. The remainder of the packet is an 802.3 Ethernet frame, followed by CRC16, padding, and end of frame (EOF) sequence. The 802.3 Ethernet frame contains the destination address, the source address, the Ethernet type that identifies the upperlayer protocol using this frame (e.g., IP), the Ethernet data containing from 64 to 1,518 bytes, and the frame check (FCK) sequence, which, like the CRC, identifies any errors in the received frame.

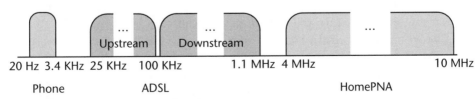

**Figure 5.5**   HomePNA spectrum allocation.

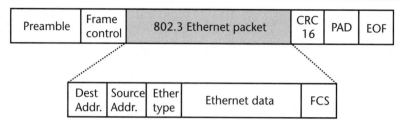

**Figure 5.6**  HomePNA frame format.

Finally, to meet latency requirements and guarantee QoS, HomePNA 2.0 provides eight priority levels and an improved collision-resolution technique that reduces the Ethernet quality problems.

## 5.4   Home Coaxial Cable Communications

In addition to a powerline and phone line, many (especially new) houses are equipped with home coaxial cable. This cable is normally used for in-home distribution of TV and radio.

The Home Cable Network Alliance (HomeCNA) is a strategic alliance that aims to standardize the physical aspects of the home coax network. The HomeCNA vision is to "leverage the preponderance of coaxial wire in the home as the networking medium for entertainment, voice and data distribution in the connected home," while its mission is to "develop, promote and proliferate a spectral application allocation standard for home coaxial wire and facilitate widespread adoption of the standard by the Cable, Telecom, Entertainment and Consumer Electronic Industries" [9]. HomeCNA has gained broader acceptance by proposing a multi-industry standard and collaborative crossendorsement of other key standards, [e.g., Versatile Home Network (VHN), UPnP, HAVi, HomeRF, CableHome].

The HomeCNA specification also proposes a frequency-allocation scheme, which enables the home coaxial network to be reused as the physical medium for higher-layer technologies by utilizing unused frequency bands. Figure 5.7 shows the proposed spectrum-allocation schema, which mixes analog and digital video signals, as well as cable modem, IEEE 1394, and Ethernet signals over the same coax. The large bandwidth capabilities of the coaxial network, along with the standard spectrum allocation, turn the coaxial network into an ideal candidate for the in-home backbone network. In Figure 5.8, a home coaxial network architecture is shown. Broadcast CATV is received by the cable modem or directly from a CATV decoder. Ethernet signals may be carried over the coaxial network to the cable modem and then via the public cable network to the Internet. IEEE 1394 signals may be

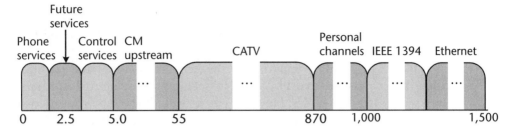

**Figure 5.7**   HomeCNA spectrum allocation.

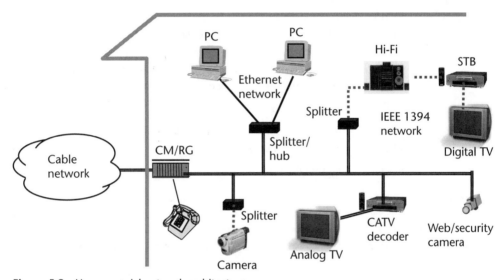

**Figure 5.8**   Home coaxial network architecture.

modulated onto the coax cable and extend the IEEE 1394 network beyond the local A/V cluster.

## 5.5   Summary

In this chapter we presented some in-home-networking technologies that have no rewiring requirements. Among them, the powerline technologies are the most common in-home network, but they suffer from noise and interference due to the powerline network/wiring environment. On the other hand, the phoneline technologies have superior performance due to their better network architecture, but they are limited by the number and location of phone jacks in the house. Although there is no critical mass of coaxial-wired houses worldwide, wherever it is available, the cable network provides a very good alternative for operating a house backbone network.

Table 5.1 provides a comparison of the technologies that reuse existing in-home wiring. We may notice that PLC and HomePlug promise up to 100 Mbps as compared with powerline cables. However, these technologies have not become widely available yet, as they are very sensitive to network architecture and load. On the other hand, HomePNA 2.0 already provides 10-Mbps capacity, and the next version aims to offer up to 100 Mbps. One would also expect that home coaxial network to provide even more than 200 Mbps. However, a large portion of the spectrum has already been allocated to legacy analog CATV, and backwards compatibility is considered mandatory.

Due to simplicity and economical reasons, technologies with no rewiring requirements are expected to acquire a large part of the existing home-networking market potential. However, it is expected that a major percentage of the in-home-network market segment will be captured by wireless technologies solutions.

**Table 5.1**  Comparison of Technologies That Reuse Existing In-Home Wiring

|  | Medium | Network | Bandwidth | QoS | Features | Applications |
|---|---|---|---|---|---|---|
| PLC | Copper | Power line | ≤20 Mbps | Not supported | Wide network availability | Home control/ automation |
| HomePlug | Copper | Power line | ≤100 Mbps | Not supported | Wide network availability | Home control/ automation, Internet |
| HomePNA 2.0 | Twisted-pair | Phone line | ≤10 Mbps (100 Mbps) | Eight priority levels | Limited only by the number and location of phone outlets | Home control/automation, Internet |
| HomeCNA | Coaxial | CATV | ≤200 Mbps | Supported | Mixes protocols in unused bands | Analog/digital TV, Ethernet, IEEE 1394 |

# References

[1]   Zahariadis, T., Pramataris, K., Zervos, N., "A Comparison of Competing Broadband In-Home Technologies," *IEE Electronics and Communications Engineering Journal (ECEJ)*, Aug. 2002, pp. 133–142.

[2]   O'Driscoll, G., *The Essential Guide to Home Networking Technologies,* Upper Saddle River, NJ: Prentice Hall, 2001.

[3]   X10, "X10 Transmission Theory," at http://www.x10.com/support/technology1.htm.

[4]   Electronics Industries Association, "EIA-600.10 Introduction to the CEBus Standard," Feb. 1995, at http://www.cebus.com/60010.doc.

[5]    Rose, B., "Home Networks: A Standards Perspective," *IEEE Communications Magazine,* Dec. 2001, pp. 78–85.

[6]    HomePlug Powerline Alliance, "The HomePlug Powerline Alliance Background Paper," at http://www.homeplug.org/whitepaper/index.html, accessed Apr. 2000.

[7]    Dhir, A., "Home Networking Using Phoneline Wiring," *Dedicated Systems Magazine,* Q2 2001, pp. 52–57

[8]    Frank, E., and Holloway, J., "Connecting the Home with a Phone Line Network Chip Set," *IEEE Micro Magazine,* 2000, pp. 1–12, at http://www.homepna.org/docs/paper500.pdf.

[9]    HomeCNA, "Home Cable Network Alliance Mission Statement," at http://www. homecna.org.

# CHAPTER 6

# Wireless In-Home Technologies

## 6.1 Introduction

In the previous chapters, we reviewed various important in-home narrowband and broadband network technologies, based on legacy (copper, twisted-pair, coaxial) and modern (plastic optical fiber) physical media. However, in all cases, the major barrier to wide deployment has been installation and maintenance costs, especially for existing houses. Pulling wires in an existing apartment is in most cases quite difficult, and most consumers are unwilling to invest in or cannot afford a large-scale home rewiring; thus, the solution is not amenable to the mass market worldwide. Finally, most existing wireline in-home networks are not flexible enough to support broadband and multimedia data applications, mainly owing to a limited number outlets and interference from other sources.

On the other hand, wireless solutions are ideal for extending the concept of home networking. The no-wires technologies are expected to be widely adopted for in-home networking and to play a key role in promoting wide acceptance of the digital house. Wireless RF and IR technologies fit better in the home environment. They provide simple and inexpensive installation, they are very flexible, and they can be gradually expanded and enlarged according to customer needs. There are two main technology categories for wireless home networks:

1. *IR:* IR transmission is a line-of-sight wireless technology. This requires that the transmitter and receiver be positioned in a direct line and that no object be positioned between them. This technology aims to interconnect appliances across short distances, normally in the same room, and does not support roaming capabilities.

2. *RF:* The RF technology is more flexible and does not require line-of-sight communication. In this way, it allows appliances to be interconnected throughout the house. RF technologies can be categorized as narrowband and spread-spectrum. Narrowband includes microwave transmission, which uses high-frequency radio waves to distances of up to 50 km, but it is not very suitable for home networks. Alternatively, spread-spectrum technology spreads the signal over a number of frequencies. This makes

the signal harder to intercept and better for the in-home environment, which is rich with reflections and mirroring effects. The most popular spread-spectrum technologies are the direct-sequence spread-spectrum (DSSS) and the frequency-hopping spread-spectrum (FHSS). DSSS systems transmit a signal over multiple frequencies simultaneously, which allows it to use the complete frequency band allocated and makes it highly resistant to interference. A receiver retrieves the original data by match filtering. The FHSS transmitters hop over entire bands of frequencies in a particular sequence, carefully selected to minimize interference on the home network, while the FHSS receivers have to hop between the same frequencies, synchronized with the transmitter. FHSS allows for a simpler radio design as compared with DSSS, but leads to more hopping overhead. DSSS and FHSS are explained in more detail in Section 6.5.

The IR line-of-sight requirement has restricted the wide deployment of IR in data networks, but assures security from eavesdropping. RF technologies on the other hand are susceptible to eavesdropping. Thus, security features and encryption techniques are normally integrated with RF home-networking products, while FHSS is more resilient to hacker attacks.

In this chapter we will describe the most widespread, mature wireless technologies that are expected to capture the maximum share of the home-networking market for different applications.

## 6.2   Infrared Data Association

For several years, IR techniques for control and data communications have been available for in-home consumer electronics usage. The best known products that use the IR technology are the TV/VCR/Hi-Fi remote controls. However, due to the strong competition in the consumer-electronics arena, most companies have tended to have their own IR standard, and although devices from the same manufacturer can communicate with each other, competing consumer-electronics products are not interoperable in most cases.

To face this interoperability problem, more than 50 manufacturers formed the Infrared Data Association (IrDA) in 1993. IrDA's objective was to specify, design, and support a ubiquitous, low-cost, PtP serial infrared (SIR) standard. In early 1994, the first IrDA standards were published. The main objective of IrDA has been to create an interoperable, low-cost IR data interconnection standard that supports a broad range of mobile appliances.

IrDA has specified the Physical and Protocol Layers necessary for any two devices that conform to the IrDA standards to detect each other and exchange data [1]. The initial IrDA 1.0 standard specified a serial, half-duplex, asynchronous communication with transfer rates from 2,400 bps up to 115,200 bps. The range between the two systems may be up to one meter with a viewing angle between 15° and 30° (see Figure 6.1). More recently, IrDA has extended the Physical Layer specification to allow data communications at transfer rates of up to 4 Mbps.

IrDA is an autoconfiguration system. To identify the devices in their communications, it uses two types of address information: the *device address* and the *connection address*. The device address is a 32-bit number that uniquely identifies an IrDA device. It is generated and maintained internally by the device. Whenever the device is initialized, it generates a 32-bit random number that will be used as its device address. If an address conflict is detected by another node, the IrDA device will be requested to change its device address, if it is in disconnected mode. The connection address is a seven-bit number that identifies a connection with a remote system. Whenever a connection is established, the primary station allocates a seven-bit random number that does not conflict with any active connection addresses and assigns it as the connection address.

### 6.2.1  IrDA Protocol Stack

The IrDA protocol stack is shown in Figure 6.2. Each IrDA-compliant device includes the core of the IrDA architecture, which consists of three standards: Infrared Physical Layer Specification (IrPHY), Infrared Link Access Protocol (IrLAP), and Infrared Link Management Protocol (IrLMP).

The IrPHY specification describes the Physical Layer of the IrDA protocol stack. The IrPHY 1.0 SIR [2] defines an infrared, asynchronous, half-duplex serial communications link for distances of at least one meter at data rates between 2,400 bps and 115.2 Kbps. The cone half-angle of the infrared transmission is at least 15°, but no more than 30°. The IrPHY 1.1 specifies SIR and fast infrared (FIR) channels and enables data rates of up to 4 Mbps.

IrLAP is located over the Physical Layer [3]. IrLAP is a high-level data link control (HDLC)–based protocol that controls access to the IR medium and provides the basic link-level connection between a pair of IrDA-compliant devices. IrLAP services may be grouped into connectionless services, which do not require a connection between the transmitter and the receiver before the communication is initiated, and connection-oriented services, in which such a connection must be established. IrLAP provides the following services:

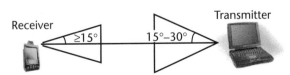

**Figure 6.1**  IrDA 1.0 viewing-angle specification.

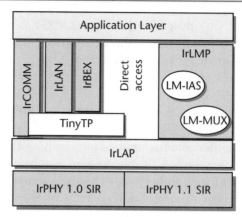

**Figure 6.2**  IrDA software architecture.

- *Discovery Service:* A connectionless service that discovers IrDA devices within communication range and available for connection. Discovery Service returns a list called a *discovery log,* which records the available devices and their descriptions.

- *Address Conflict Service:* A connectionless service that is used to resolve device address conflicts. Within the discovery operation, if two devices are found to have the same device address, Address Conflict Service of the conflicting devices is invoked to select new nonconflicting device addresses.

- *Sniffing Service:* A connection-oriented service that is used to indicate if a device is available for a new connection or not. A device that is available for a new connection enters a special mode, called *sniffing.* During the discovery service, if a device is in this mode, it sets a specific flag (Sniff flag) into the discovery log indicating its availability, and attaches its QoS parameters (e.g., bandwidth, processing power, current load). A sniff request can be canceled by issuing a request primitive with the Cancel flag set to true.

- *Connect Service:* A connectionless service that establishes a connection between two devices with a specific QoS. The device that aims to set up a connection initiates the discovery service, finds the sniffing devices, and then requests of the sniffing device that a connection be established. After a connection has been established, a connection handle is returned, which is used for future references to the connection.

- *Data Service:* A connection-oriented service that sends data to the receiver. According to one flag (Expedited-Unreliable flag), the data sent may be either reliable, error-free, and in the proper order (e.g., data communications) or unreliable, expedited, and unsequenced (e.g., control messages). In both cases no confirmation is returned to the sender.

- *Unit Data Service:* A connectionless service that provides an unreliable way to transmit data outside of a connection. Data is broadcasted and cannot be directed to a specific device address.

- *Status Service:* A connection-oriented service that informs the upper layer about the quality of the link. It is invoked either when the link is experiencing high levels of noise or when all connection activity has ceased. It does not affect the transmission of the data, but provides a flow mechanism for the upper layer.

- *Reset Service:* A connection-oriented service that causes all unacknowledged data units to be discarded. All counters and timers are reset. A reset only occurs if both ends of the connection agree to it.

- *Disconnection Service:* A connection-oriented service that terminates the logical connection, and all outstanding data units are discarded. No confirm primitive is needed since the disconnect is always successful.

The IrLMP Layer is located over the IrLAP Layer [4]. IrLMP has two primary functions: it is the link management multiplexer (LM-MUX), which enables multiple entities within any pair of IrDA devices to use the same single IrLAP connection simultaneously and independently, and the link management information access service (LM-IAS), which discovers and registers the services that are offered by a peer device.

Apart from the base standards, IrDA specifies TinyTP, a transport protocol (TP) that is much more lightweight than TCP, designed to provide application-level flow control and SAR of application data units [5].

Over the TinyTP, IrDA specifies the Infrared Communication (IrCOMM) and the Infrared Local Area Network Protocols (IrLAN) [6]. The IrCOMM Serial and Parallel Port Emulation Protocol enables the redirection of conventional serial and parallel ports over the IR medium, allowing IrDA links to operate as cable replacements. Similarly IrLAN provides wireless access to LAN. IrLAN was developed to allow an IrDA-enabled device to access a LAN over the IR medium.

In 2001, IrDA also specified a protocol for generic IR object exchange, called *Infrared Object Exchange Protocol* (IrOBEX). IrOBEX is an industry standard, which defines how objects can be shared and exchanged between different IrDA devices, thereby enabling rapid application development and interaction with a broad range of devices, including PCs, PDAs, data collectors, cellular phones, handheld scanners, and cameras. IrOBEX is based on the Hypertext Transfer Protocol (HTTP), but is more compact. It piggybacks objects within IrDA communication transactions, enabling the transmitted objects to be recognized and handled intelligently on the receiving side. In this way, the application developer does not have to worry about the low-level IrDA functions of link discovery, setup, and maintenance, but can instead focus on higher-level application development. Finally,

IrOBEX enables exchange of data-object information, such as object description headers, along with the data object itself.

### 6.2.2  IrDA Communication Range

The initial IrDA 1.0 specification targeted short distances of up to 1m. However, in many cases longer link distances are required. To increase the link distance, either the transmitted light intensity, the receiver sensitivity, or both must be increased. In case of communication with a standard IrDA end-point device, the peer end-point has to increase both the transmitter intensity and the receiver sensitivity. The problem with this expansion is that intensity drops with the square of the distance. For example, extending the link distance from 1m to 3m requires nine times the initial power on the transmitter or nine times the sensitivity on the receiver, while maintaining the dynamic range, so that the device will still work at 1m. To achieve this enhancement, one solution is to use laser diodes, but they are more expensive, and they are also hazardous to the eyes if they are stronger than 1 mW. Better solutions that use lenses to focus the IR beam have also been proposed.

The IrDA standard provides a remarkable, low-cost, high-bandwidth solution. However, like all IR solutions, it is expected to have limited deployment in home data networks, due to its line-of-sight requirement.

## 6.3  Digital Enhanced Cordless Telecommunications

Digital Enhanced Cordless Telecommunications (DECT) [7] is a flexible digital radio access standard for cordless communications in residential, corporate, and public environments. It operates in the preferred 1,880- to 1,900-MHz band, using Gaussian frequency shift keying (GFSK) modulation, in Europe, while in other parts of the world, frequencies between 1,900 and 1,930 MHz are used.

DECT was introduced in the early 1990s by ETSI for cordless telecommunications, mainly focusing on telephony and ISDN access [8]. After the first edition of the DECT standard in 1992, the DECT standardization work concentrated on the definition of standard interworking profiles. The Generic Access Profile (GAP), the first profile, was completed in 1994. GAP is the basis for all other DECT speech profiles and contains the protocol subset required for basic telephony service in residential cordless telephones, business wireless PABX, and public-access applications. Other interworking profiles, such as the DECT/GSM Interworking Profile (GIP), the DECT/ISDN Interworking Profile (IIP), the DECT/Radio Local Loop Access Profile (RAP), the DECT/Cordless Terminal Mobility (CTM) Access Profile (CAP), and Multiple Data Service Profiles, followed. The extensions and enhancements to the DECT base standard led to DECT's second edition in 1995. Some examples of these extensions are inclusion of emergency call procedures to aid acceptance of DECT for public-access applications, definition of the wireless relay station (WRS) as a new system component to enable more cost-efficient infrastructures, and description of the optional direct portable-to-portable communication feature for DECT.

Currently, DECT is supported and promoted by the DECT Forum, with representatives in all of the major geographical regions around the world. DECT contains many forward-looking technical features and profiles. Its multicarrier (MC)/TDMA/time division duplex (TDD) radio access method and continuous dynamic channel selection and allocation (DCSA) capability enable the use of high-capacity picocellular systems even in busy or hostile radio environments. These methods enable DECT to offer excellent QoS without the need for frequency planning and allow it to be used in a variety of scenarios, including the following:

- Cordless telephony;
- Cordless network access for a generic voice or data network;
- Internet access;
- WLL;
- Cordless terminal mobility;
- Data services, circuit- or packet-switched links;
- Multimedia access;
- IMT-2000 services for fixed and low-mobility users;
- Narrowband wireless local area network (WLAN).

### 6.3.1  DECT Network Architecture

A reference DECT network architecture is based on small cells, called *picocells*, as Figure 6.3 shows. DECT terminals are classified as fixed points (FPs) using one or more base stations [radio fixed point, (RFP)] and portable points (PPs), which are

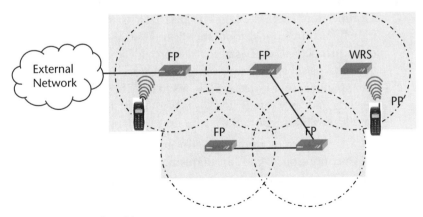

**Figure 6.3**  DECT network architecture.

the cordless terminals. According to the number and type of RFP, each FP covers a specific area (picocell). One or more PPs may move within the cell and synchronize their transmissions according to the time base defined by the FP.

To cover larger areas, more FPs can be interconnected with wireline links. Moreover, WRSs allow the extension of an FP's radio range. The WRS acts as a terminal, when seen from the FP's point of view, and as a base station when seen from the PP's. Additionally, DECT supports ad hoc networking. Direct communication between two PPs is allowed, but one of them should act in this case as FP.

It is important to note that the DECT standard covers only the so-called air interface between a FP and a PP. In this way, the base station operates as a toolbox with protocols and messages. The FP and the PP can make selections (profiles) to access any specific type of external network. In addition to cordless operation, DECT makes network-specific services and features (e.g., mobility) available to the user transparently through the DECT common air interface.

Finally, the DECT standard does not set a limit on the number of base stations and cordless terminals. Infrastructures using the DECT technology can theoretically support traffic densities of up to 10,000 Erlang/km$^2$, which is comparable to 100,000 voice calls in office environments.

### 6.3.2  DECT Operational Characteristics and Features

DECT systems' efficiency is based on the following mechanisms and operational characteristics:

- *MC/TDMA/TDD Transmission:* The DECT radio interface is based on the MC/TDMA/TDD radio access methodology. Basic DECT frequency allocation uses 10 carrier frequencies in the 1,880 to 1,900-MHz range. The time is organized in frames. The duration of each frame is 10 ms and consists of 24 time slots (Figure 6.4).

  To simplify implementation, the time slot is separated in two fixed parts of 12 time slots each (TDD). The first part is used for FP-to-PP transmissions (downlink) and the second for PP-to-FP transmissions (uplink). Due to this radio protocol, DECT is able to offer widely varying bandwidths by combining multiple channels into a single bearer. For data-transmission purposes, throughput rates of up to 552 Kbps in 24-Kbps steps can be achieved with full security and very low BER ($10^{-9}$). With eight-level modulation, the maximum data rate (unidirectional) may be up to 2 Mbps.

- *DCSA.* To increase frequency-allocation efficiency, capacity, and QoS, DECT uses a continuous DCSA mechanism. The DCSA mechanism aims to set up the radio links on the channel with the least interference available. To achieve this, DECT systems have to monitor the RF signal strength on all idle channels at least once every 30 seconds. In this way, each PP can be attached to the base

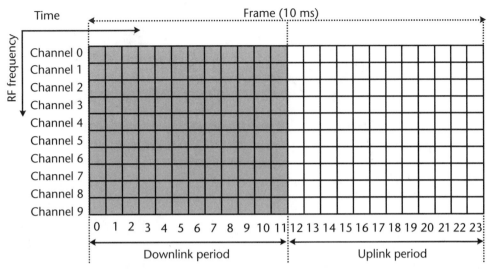

**Figure 6.4**    DECT time-frame structure.

station with a stronger RF signal, given that it has the required access rights and selects the optimal channels to set up a new communication link.

• *Mobility and Handover*. Mobility functions in the DECT architecture provide the PP with the ability to roam freely through a (multi)cellular residential or business infrastructure. Although networkwide mobility is outside the scope of the DECT standard, wireless users with authorized access to the network can initiate and receive calls at any location within the DECT coverage area and roam between DECT picocells when in active communication. When the radio channel is interfered with, the handover capability of DECT assures transparent hopping to a newly selected radio channel that is not experiencing interference. Figure 6.5 shows the intracell handover case, where a terminal, experiencing interference on one radio channel (step 1), selects a new channel and establishes a new connection (step 2), and then releases the initial channel (step 3).

    Figure 6.6 shows the intercell handover case, where the terminal changes base stations. During the handover period, both radio links are active in parallel with identical speech or data information, or both. After some time, the base station analyzes the quality of the links, determines which radio link is experiencing less interference, and releases the other link.

• *Diversity:* To deal with situations in which the channel fades very quickly and handover is not sufficient, DECT base stations can be equipped with antenna diversity. In this case, the PP uses a special signaling protocol to control FP antenna diversity. Due to the symmetric TDD radio link between the FP and

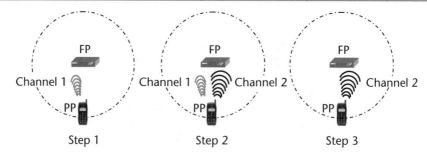

**Figure 6.5**   Intracell handover.

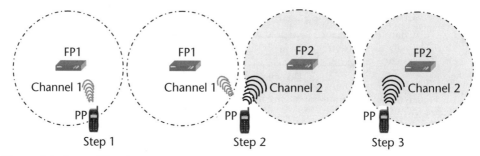

**Figure 6.6**   Intercell handover.

PP, the FP antenna diversity improves both the uplink quality and the down-link quality at low speeds.

- *Security:* DECT provides a security mechanism sufficient for all in-home communications. The DCSA mechanism is combined with effective subscription and authentication protocols to prevent unauthorized access. Moreover an advanced ciphering concept provides protection against eavesdropping.

- *Spectrum Efficiency:* In Europe, DECT uses 10 carriers in the 1,880- to 1,900-MHz frequency range. In other parts of the world, frequencies between 1,900 and 1,930 MHz may be used. Moreover, the North American Personal Wireless Telecommunications standards Personal Wireless Telecommunications (PWT) and PWT/Enhanced (PWT/E) (TIA) are based on DECT. PWT operates in the U.S. unlicensed band at 1,910 to 1,920 MHz, while PWT/E is an extension into the licensed bands at 1,850 to 1,910 MHz and 1,930 to 1,990 MHz.

### 6.3.3   DECT in the Wireless Home Network

DECT is the default standard for cordless phone communications for in-home and corporate environments. With regard to the WLAN application, DECT is based on the Data Packet Radio Service (DPRS), which defines features and services common to all packet-data applications, with and without mobility. Wireless data communications using DECT technology have all of DECT's advantages, including the efficient use of the spectrum, robust communication due to the DCSA, mobility and handover, authentication, and data encryption. Due to wide market penetration, DECT terminals are also comparatively cheap.

Its main disadvantage is its relatively low data rate (552 Kbps). This can be enhanced with higher-level modulation formats (up to 2 Mbps), which is adequate for many in-home applications, but not enough for multimedia applications.

## 6.4  Bluetooth

Bluetooth serves as a universal low-cost, user-friendly air interface that aims to replace the plethora of proprietary interconnect cables between various personal devices [9, 10]. Bluetooth is a short-range (10 cm–10m) FHSS wireless system providing up to 1 Mbps in the unlicensed 2.4-GHz industry, scientific, and medical (ISM) band. There are also efforts to extend the range of Bluetooth with higher-power devices spanning longer distances.

The Bluetooth wireless technology supports both PtP and PtMP connections. Currently up to seven slave devices can communicate with a master radio in one device. It also provides for several small wireless networks, called *piconets*, to be linked together in ad hoc networking mode, allowing for extremely flexible configurations such as might be required for meetings and conferences.

The technology was named Bluetooth to honor the Danish king Harald Blåtand, who united Demark and Norway in the tenth century, much as wireless technology unites people today. The name Blåtand translates loosely to "Bluetooth" [11].

Bluetooth technology has been developed and promoted by the Bluetooth Special Interest Group (BSIG), an industry-based association. The founding members of the group were IBM, Intel, Ericsson, Nokia, and Toshiba. Since 1999, 3Com, Lucent, Microsoft, and Motorola have joined, while by December 2000, BSIG had more than 2,000 member companies. The Bluetooth standard published by BSIG is open. It includes an air-interface specification, a host-controller interface that specifies the interface between a host and a Bluetooth device, and interoperability profiles, which assure interoperability between Bluetooth equipment from different vendors.

### 6.4.1  Bluetooth Network Architecture

Within the Bluetooth specification, communication is based on ad hoc networking. That means that there is no base station or access point, but the standard defines mechanisms and messages for Bluetooth devices to discover each other and establish communication links. A standard example of ad hoc networking is where a number of people are around a conference table and information is shared between them. A more private operational example is where Bluetooth devices communicate in a "closed" group. For example, when the mobile phone, laptop, and PDA of the same person communicate, but no other devices within range are allowed to communicate by any means. Nevertheless, after initial communication and without the need for base stations, the Bluetooth standard defines a structured network architecture based on a star network topology, called a *piconet*. A piconet is a very short-range wireless network with a diameter up to 10m. When a Bluetooth device initiates a communication, it defines a new piconet cell, with the device as the center of the cell. This device is the master Bluetooth device of this cell, and all other devices in this cell are considered slave devices. Due to a limitation of the air interface, which uses a three-bit Identity field, one master device may communicate with up to seven active slave devices. If more slave devices are located within the same piconet, they remain inactive ("parked") and no resources are allocated to them. During the existence of a piconet, the master device may be changed dynamically.

Piconet cells may be combined to form a "scatternet." In a scatternet, one or more Bluetooth devices are members of more than one piconet, acting as the master device in one cell and a slave device in another. Any Bluetooth device may switch between master and slave modes, but cannot be in both master and slave mode at the same time.

Figure 6.7 shows a Bluetooth network topology. A scatternet consists of eight Bluetooth devices, distributed in two piconets. Piconet 1 consists of one master Bluetooth device and four slave devices, while Piconet 2 consists of one master and three slave devices. Moreover the master device of Piconet 2 operates as an additional slave device in Piconet 1.

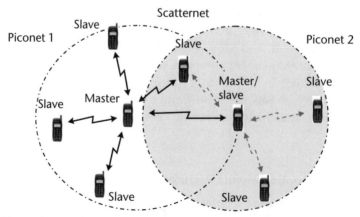

**Figure 6.7** Bluetooth network architecture.

### 6.4.2 Air Interface Format

Bluetooth operates in the unlicensed frequency range of 2.4 to 2.4835 GHz, with channel spacing of 1 MHz. Bluetooth is an FHSS system with 79 channels, a hopping rate of 1,600 hops per second, and a different hopping sequence per piconet.

Frequency hopping is used to minimize the interference effects from other users or other protocols in the unlicensed band. The clock of the master device defines the hopping events in a piconet, and all slaves of the piconet have to synchronize with this clock. Apart from the frequency hopping, the master clock is used for the time division slot and for encryption. Bluetooth uses a slotted TDD system. The TDD slot structure is shown in Figure 6.8. Each packet is transmitted on a different frequency, normally within a single time slot, but can be extended to cover up to five slots. If more than one slot is used for a single packet transmission, the frequency does not hop until the entire packet has been transmitted. Each time slot has a duration of $625\,\mu\mathrm{sec}$, with a $220\,\mu\mathrm{sec}$ guard time between the end of the reception of one packet and the start of the transmission of the next packet, guarantying against packets collisions.

The Bluetooth packet structure is shown in Figure 6.9. It consists of an access code, a packet header, and a payload [12]. The access code is used for identification and packet synchronization. The Bluetooth standard defines a number of access codes: the channel access code (CAC), which is derived by the master device and identifies the piconet cell; the device access code (DAC), which is derived from the slave devices and identifies the devices; and the inquiry access code (IAC), which is used for device inquiries. In all cases, the synchronization word (pattern) is included or can be calculated. The Bluetooth packet header contains the address of the active piconet member, the packet type that determines whether the packet belongs in an asynchronous connectionless (ACL) bearer or in a synchronous connection-oriented (SCO) bearer, the flow-control information used for flow control of ACL connections, the error-control (EC) acknowledgement bit (ACK = 1, NACK = 0),

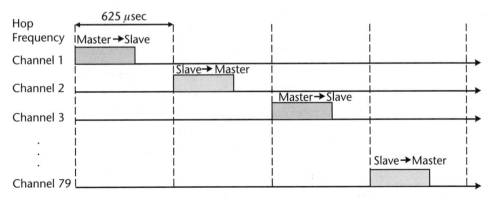

**Figure 6.8** Bluetooth FSS TDD slot structure.

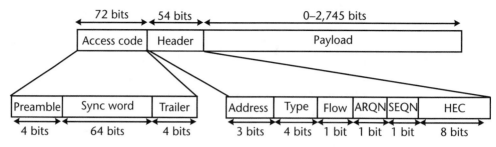

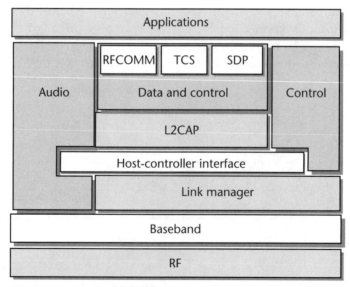

Figure 6.9   Bluetooth packet structure.

the sequence number as a modulo 2 sequential packet number, and the header error control (HEC). The header is encoded with 1/3 FEC code, resulting in 54 bits (= 3 × 18 bits).

### 6.4.3   Bluetooth Protocol Stack

The Bluetooth protocol stack architecture is shown in Figure 6.10. It is a layered stack that supports physical separation between the link manager and the higher layers at the host control interface (HCI), which is common in most Bluetooth implementations. The Baseband Layer provides the lower-layer functionality, which is required for air interface packet framing, establishment and maintenance of piconets, and link control.

Figure 6.10   Bluetooth software architecture.

The link manager is responsible for link setup and control and supports a number of procedures, including authentication, encryption control, physical parameter control, master-to-slave switching, and the like. The HCI provides for a mechanism whereby the higher layers of the protocol stack can delegate the decision as to whether to accept connections to the link manager and whether to switch on filters at the link manager. It supports bidirectional transmission of connectionless ACL data and connection-oriented SCO streams, like audio, across it.

The Logical Link Control Adaptation Layer Protocol (L2CAP) provides connection-oriented and connectionless data services to higher-layer protocols. It supports multiplexing of higher-layer protocols, establishment and removal of logical connections for SCO services, SAR to allow transport of packets of up to 64 KB, and the like. It also enables the negotiation of the connector's QoS parameters.

The Service Discovery Protocol (SDP) is a higher-layer Bluetooth-specific protocol, which allows Bluetooth devices to discover what services are available on a device. SDP follows the client-server paradigm. It uses the service discovery database server to store and retrieve service records and attributes. RF communication (RFCOMM) provides an emulation of serial ports. It is based on the ETSI GSM mobile-telephone specification, and enables two Bluetooth devices to communicate as if they where connected via an RS-232 connection. RFCOMM may emulate up to 64 serial ports between two Bluetooth devices.

The Telephony Control Specification (TCS) provides an Adaptation Layer that enables Q.931 call control services over L2CAP.

### 6.4.4   Example Bluetooth Products

To be attractive to the mass market, the Bluetooth specification had to be comparable as far as maintenance, user-friendliness, and cost to serial cable systems [13]. The cost-competitive requirement has led to the point that Bluetooth chips should cost less than $5. Currently, this target has not yet been reached; nevertheless advances in very large scale integration (VLSI) have shown that this projection is be reasonable. Figure 6.11 shows a number of Bluetooth modules and their remarkably small size.

The small size and weight of the Bluetooth chipsets have released the system designers from many constraints and have led to some very exciting and easy-to-use products. Bluetooth may be included in practically in any device. Some examples include the following:

- *Wireless Hands-Free Headsets:* The user can initiate or answer a call with voice commands. In Figure 6.12, a number of Bluetooth handsets are shown. Their weight, including Bluetooth chipset and battery, is around 10g and their standby time may be up to 48 hours.

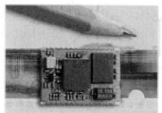

Tochini-Mitsumi/preassembly module          Ericsson/Bluetooth kit

**Figure 6.11**   Bluetooth modules.

- *Cable-Free Networking:* USB to Bluetooth adapter and Bluetooth PC cards fall into this category. Figure 6.13 shows representative products of this category. Figure 6.14 shows Bluetooth car kits.
- *Human Interface Devices:* This category is maybe the most impressive. For the sake of example, we refer to the Ericsson/Anoto ChatPen. Similar products may be available from other vendors. The pen, together with a specially patterned paper, enables the user to store and wirelessly transmit anything he writes or draws. As Figure 6.15 shows, the pen, in addition to an ink cartridge, includes a digital camera, an image-processing unit, and a Bluetooth radio transceiver.

The Ericsson/Anoto ChatPen is like an ordinary ballpoint pen. However, as the user writes, the dots of the ink are illuminated by infrared light, and the digital camera records 100 snapshots of the pattern per second. The image processor calculates, in real-time, the exact position in the entire proprietary pattern and information about how the pen is held is also gathered and stored. The information is transmitted by the Bluetooth transceiver, either directly to a computer, or forwarded via a relay device (e.g., mobile phone or handheld device) to a server.

## 6.5   IEEE 802.11/IEEE 802.11b

IEEE 802.11 has been one of the most important pioneering efforts to specify and develop a data-oriented WLAN standard [14, 15]. The widespread acceptance of

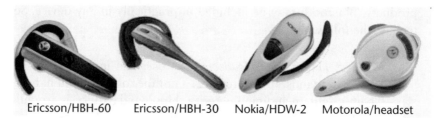

Ericsson/HBH-60      Ericsson/HBH-30      Nokia/HDW-2      Motorola/headset

**Figure 6.12**   Bluetooth handsets.

Mitsumi/USB adapter     D-Link/USB adapter          Anycom/USB adapter

**Figure 6.13**   Bluetooth adapters and devices.

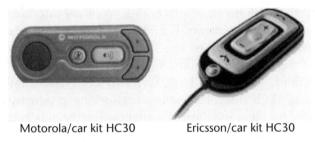

Motorola/car kit HC30          Ericsson/car kit HC30

**Figure 6.14**   Bluetooth car hands-free adapters.

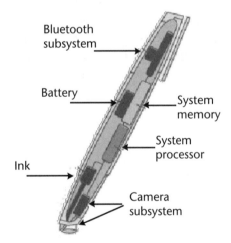

**Figure 6.15**   Ericsson/Anoto Bluetooth ChatPen.

IEEE 802.11/802.11b derives from industry standardization, which has ensured product compatibility and reliability among the various manufacturers. IEEE 802.11 has been so widely accepted worldwide that the general term *wireless LAN* or *WLAN* refers to the IEEE 802.11 standard or its evolution, IEEE 802.11b.

Definition of IEEE 802.11 started in 1990, and the first version of the standard was almost finalized by mid-1996. In 1997, the Institute of Electrical and Electronics Engineers (IEEE) validated the initial 802.11 specification as the standard for WLANs. The first version of 802.11 provided for 1–2-Mbps data rates and a set of fundamental signaling methods and services. The major drawback of this version was limited throughput.

In 1999, recognizing the critical need for higher data-transmission rates in most business applications, the IEEE enhanced the 802.11 standard with the 802.11b standard, which is able to support transmissions of up to 11 Mbps. IEEE 802.11b enables WLANs to achieve performance and throughput comparable to wired 10-Mbps Ethernet.

In parallel with the standards bodies, to ensure interoperability and compatibility across all market segments, IEEE 802.11 product manufactures have agreed on a compliance procedure called the *Wireless Fidelity Standard* (Wi-Fi). Moreover, a Wireless Ethernet Compatibility Alliance (WECA) has been formed to certify Wi-Fi interoperability of new products, to certify crossvendor interoperability and compatibility of IEEE 802.11b wireless-networking products, and to promote IEEE 802.11b for business and the home applications. Members include WLAN semiconductor manufacturers, WLAN providers, computer-system vendors, and software makers—such as 3Com, Aironet, Apple, Breezecom, Cabletron, Compaq, Dell, Fujitsu, IBM, Intersil, Lucent Technologies, AVAYA, Agere, No Wires Needed, Nokia, Samsung, Symbol Technologies, Wayport, and Zoom.

IEEE 802.11 and IEEE 802.11b have been tested and deployed for years in corporate, enterprise, private, and public environments (e.g., hot-spot areas) and are among the major candidate solutions for home networking. The IEEE 802.11 standard supports several WLAN technologies in the unlicensed bands of 2.4 and 5 GHz and share the same MAC over two Physical Layer specifications: DSSS and FHSS technologies. IR technology is also supported, but has not really been adopted by any manufacturer.

### 6.5.1   Network Architecture

Wireless 802.11 end stations can be PC-based terminals equipped with an 802.11 PC card (PCMCIA), a peripheral component interconnect (PCI), or an industry standard architecture (ISA) network interface card (NIC) or with embedded solutions in non-PC clients (e.g., an 802.11-based PDA or handset).

IEEE 802.11 supports communication of terminals via two general network architecture topologies: *structured* mode and *ad hoc* mode. In the structured mode (Figure 6.16), the terminals communicate with the backbone network via an access point. The access point acts as the base station for the wireless network, aggregating access for multiple wireless stations onto the wired network. The configuration, which consists of at least one access point connected to the wired network

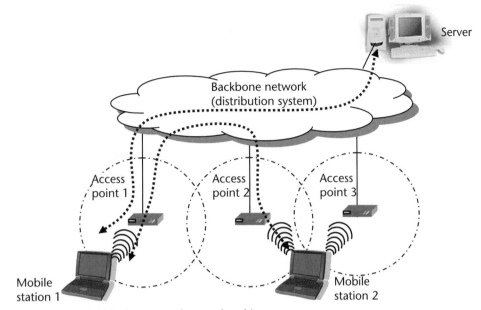

**Figure 6.16**   IEEE 802.11 structured network architecture.

infrastructure and a set of wireless end stations, is called a *basic service set* (BSS). Multiple access points are normally connected via a fixed, wired network (i.e., Ethernet). Wireless connections between the access points are supported via a special frame format that effectively tunnels original frames over the 802.11 wireless networks. A set of two or more BSSs is called an *extended service set* (ESS).

The structured topology is useful for providing wireless convergence of buildings, campuses, or hot-spot areas by developing multiple access points whose radio coverage overlaps to provide complete coverage.

The ad hoc networking topology allows direct communication between the terminals without the need for an access point (Figure 6.17). Ad hoc networking is useful for fast and easy setup of a wireless network anywhere for file-sharing applications between the participants. Common location examples may include a hotel room, a convention center, an airport, or a client site. The MAC Protocol of the standard allows both types of topologies to coexist in the same network.

### 6.5.2   Protocol Stack

Like all IEEE 802.x protocols, IEEE 802.11 covers the lower layers of the OSI model and specifies the Physical and the MAC Protocol [16]. Moreover, the IEEE 802.2 logical link control (LLC), 48-bit addressing, and upper protocol stack layers remain unchanged, as with other 802 LANs, allowing for very simple bridging from wireless to IEEE wired networks (Figure 6.18).

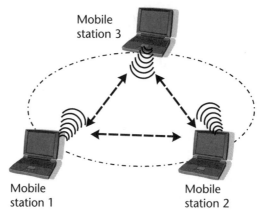

**Figure 6.17**   IEEE 802.11 ad hoc network architecture.

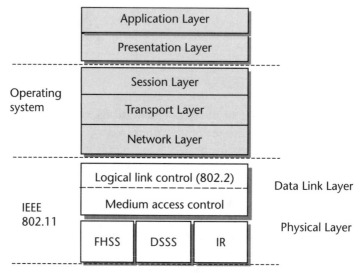

**Figure 6.18**   IEEE 802.11 protocol stack.

### 6.5.2.1   802.11 Physical Layer

The 802.11 standard defines three types of Physical Layer specifications: DSSS, FHSS, and IR. In practice, only the first two, DSSS and FHSS, have presence in the market.

Just like Bluetooth, the radio-based IEEE 802.11 standards operate within the unlicensed 2.4-GHz ISM band. This frequency band is recognized by international regulatory agencies, such as the FCC (United States), ETSI (Europe), and the MKK (Japan), for unlicensed radio operations. Table 6.1 summarizes the worldwide frequency allocations for unlicensed operation in the ISM band.

In the DSSS technique, the 2.4-GHz band is divided into 14 channels of 22 MHz each. Adjacent channels partially overlap; only three are completely nonoverlapping. Data is sent across one of the channels without hopping to other channels. The user data is modulated by a single predefined wideband-spreading signal. The receiver knows this signal and is able to recover the original data.

To compensate for noise on a given channel, 802.11 DSSS uses a technique called *chipping*. Each data bit is converted into a series of redundant bit patterns called *chips*. The transmitter encodes with an exclusive-OR (XOR) gate all data sent via an 11-bit, high-speed, pseudorandom numerical (PRN) sequence called the *Barker sequence* (Figure 6.19).

The term *chip* is used instead of *bit* to indicate that the Barker sequence itself does not carry any binary information. Each 11-chip sequence represents a single data bit (1 or 0) and is converted to a waveform, called a *symbol*, which is sent over the air. The effect of the Barker sequence is to spread the transmitted bandwidth of the resulting signal by a ratio of 11:1. After encoding, the symbols are transmitted at a rate of 1 Msps (megasymbols per second), using a modulation called *binary phase shift keying* (BPSK), which achieves 1 Mbps, or using a more sophisticated modulation called *quadrature phase shift keying* (QPSK), which achieves a data rate of up to 2 Mbps. The inherent redundancy of each chip combined with spreading the signal across the 22-MHz channel provides a form of error checking and correction. Even if part of the signal is damaged, it may still be recovered in many cases, minimizing the need for retransmissions.

Using the FHSS technique, the 2.4-GHz band is divided into a large number of subchannels. The number of subchannels differs between geographical regions (i.e., 79 frequencies in United States and Europe and 23 in Japan). The peer communication end-points agree on the frequency-hopping pattern, and data is sent over a sequence of the subchannels. The transmitter sends data over a subcannel for a fixed length of time, called the *dwell time*, then changes frequency according to the hopping sequence and continues transmission in the new frequency. Each conversation within the 802.11 network occurs over a different hopping pattern, and the patterns are designed to minimize the possibility of two senders using the same subchannel simultaneously. As the dwell time is rather long, the transmitter can send multiple consecutive symbols at the same frequency. FHSS techniques allow for relatively simple radio design, but are limited to speeds no greater than 2 Mbps. This

**Table 6.1**   Global Spectrum Allocation at 2.4 GHz

| Region | Allocated Spectrum (GHz) |
|---|---|
| United States | 2.4000–2.4835 |
| Europe | 2.4000–2.4835 |
| Japan | 2.471–2.497 |
| France | 2.4465–2.4835 |

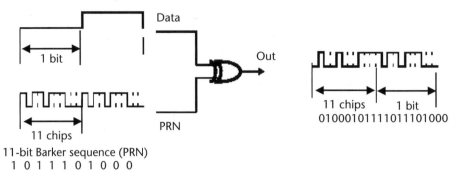

**Figure 6.19**   802.11 DSSS encoding using the Barker sequence.

limitation is driven primarily by FCC regulations that restrict subchannel bandwidth to 1 MHz, leading to high hopping overhead.

### 6.5.2.2   802.11b Physical Layer Enhancements

The main enhancement of 802.11b was the standardization of a Physical Layer able
to support higher speeds of 5.5 and 11 Mbps. As 802.11 FHSS systems cannot support the higher speeds without violating current FCC regulations, the DSSS technique was selected as the exclusive Physical Layer technique. In this way, 802.11b is
backwards compatible and can interoperate at 1 Mbps and 2 Mbps only with the
802.11 DSSS systems and not with FHSS systems.

To increase the data rate, rather than the 11-bit Barker sequence, 802.11b specifies an advanced coding technique called *complementary code keying* (CCK). CCK
consists of a set of 64 code words, each eight bits long, which can be distinguished by
the receiver even in the presence of noise or interference. The CCK encodes four bits
per carrier to achieve a data rate of 5.5 Mbps, and eight bits per carrier to achieve 11
Mbps. Both speeds use QPSK modulation and a 1.375-Msps symbol rate. Table 6.2
summarizes the differences between the 802.11 Physical Layers.

The 802.11b standard provides rates of 11 Mbps across distances of 300m to
400m in open, outdoor environments and 30m to 50m in indoor environments with
low noise. To support noisy environments as well as extended range, 802.11b
uses dynamic rate degradation. When the terminal moves beyond optimal range or

**Table 6.2**   802.11 Physical Layer Differences

| Physical Layer | Data Rate (Mbps) | Bits per symbol | Code | Modulation | Symbol Rate(Msps) |
|---|---|---|---|---|---|
| 802.11 | 1 | 1 | Barker sequence | BPSK | 1 |
| 802.11 | 2 | 2 | Barker sequence | QPSK | 1 |
| 802.11b | 5.5 | 4 | CCK | QPSK | 1.375 |
| 802.11b | 11 | 8 | CCK | QPSK | 1.375 |

if substantial interference emerges, 802.11b degrades transmission to lower speeds, falling back to 5.5, 2, and finally 1 Mbps. If the terminal reenters the optimal range, or the source of interference disappears, the connection will automatically accelerate.

### 6.5.2.3   802.11 MAC Layer

IEEE 802.11 specifies a single MAC Protocol for all 802.11 Physical Layers. This design decision is quite important as it greatly simplifies interoperability, while enabling chip vendors to achieve higher production volumes and keep prices low.

The 802.11 MAC, like the 802.3 Ethernet MAC, has to support multiple users on a shared medium by having the sender sense the medium before accessing it and by handling collisions that occur when two or more terminals try to communicate simultaneously. In the Carrier Sense Multiple Access with Collision Detection Protocol (CSMA/CD) used in the wired Ethernet, a terminal must be able to transmit and listen at the same time. However, in radio systems the terminal is not able to transmit and receive simultaneously and, thus, is not able to detect a collision. Therefore, 802.11 uses a modified protocol, called *Carrier Sense Multiple Access with Collision Avoidance* (CSMA/CA) or *Distributed Coordination Function* (DCF). CSMA/CA (Figure 6.20) attempts to avoid collisions by using explicit packet acknowledgment (ACK), which means that an ACK packet is sent by the receiving station to confirm that the data packet arrived intact.

Initially, a terminal senses the medium. If the medium is idle, the terminal may transmit; if it is busy, the terminal has to wait. After the medium becomes idle, the terminal has to wait two additional time periods. The first period depends on the packet to be transmitted. If it is an ACK, this period is one short interframe space (SIFS); otherwise, it is a DCF interframe space (DIFS). The second is a random back-off period, which prevents multiple terminals from seizing the medium immediately after completion of the preceding transmission. After both periods have passed and the channel has not already been seized, the terminal may start transmission. Otherwise, the whole process restarts.

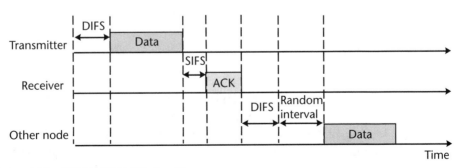

**Figure 6.20**   802.11 CSMA/CA algorithm.

To ensure robustness, the 802.11 MAC Layer provides CRC and packet fragmentation. To ensure that a packet has not been corrupted, the transmitter calculates and attaches a 32-bit CRC checksum to each packet before it is transmitted. The receiver recalculates the CRC, and if it matches, the packet has not been corrupted. Packet fragmentation splits large packets into smaller units, since the larger packets have a higher probability of being corrupted. This technique reduces the number of retransmissions or requires retransmission of shorter messages; thus, it improves the overall performance of the wireless network.

### 6.5.3   Security

Like all RF wireless protocols, IEEE 802.11 is sensitive to eavesdropping. To avoid malicious access, IEEE 802.11 provides for both MAC Layer access control and encryption mechanisms, known as wired-equivalent privacy (WEP). WEP aims to provide security equivalent to that of wired counterparts.

For the access control, an extended service set identifier (ESSID), also known as a WLAN service area ID, is programmed into each access point. Each client that aims to attach or communicate with this access point has to know the ESSID. Moreover, the access point has a table of MAC addresses called an *access control list,* and access is enabled only for the terminals whose MAC addresses are on the list.

The standard provides for data encryption using a 40-bit shared-key RC4 pseudorandom number generator (PRNG) algorithm. Optionally, some vendors provide 64-bit or 128-bit shared keys. Moreover, when encryption is activated, the access point will send an encrypted challenge packet to potential clients. The clients must use their keys to encrypt the correct response to authenticate themselves and gain network access.

Much attention has been paid recently to the fact that WEP is not a so-called industrial-strength encryption protocol [17]. While WEP does not meet the needs of all applications, for many deployments

> WEP has been, and continues to be, a very effective deterrent against the vast majority of attackers who might attempt to compromise the privacy of a wireless LAN. WEP is Wired Equivalent Privacy, and just as with a wired network, security layers can be deployed above the wireless LAN layer. An example here is the use of VPN, which provides end-to-end security, for which the wireless LAN is transparent [18].

IEEE 802.11b WLANs are already broadly used in several large vertical markets. IEEE 802.11b provides for flexible, robust, and reliable 11-Mbps performance and interoperable solutions from multiple vendors. Many vendors worldwide provide IEEE 802.11b products. Most of them target laptop users; thus, they provide IEEE 802.11b cards with a PC card (PCMCIA) interface. A small selection of IEEE 802.11b products is shown in Figure 6.21. Special mention should be given to the

OriNOCO    SpeedStream 802.11b    LinkSys WPC11    3Com Wireless LAN

**Figure 6.21**    A selection of IEEE 802.11b products.

NOKIA D211/D311 card (Figure 6.22), which integrates the IEEE 802.11b Protocol with cellular (GPRS) technology.

The major drawback to the standard is the lack of QoS and synchronous channels for voice and video communications. Therefore, a number of IEEE 802.11 extensions have been initiated aiming to enhance the bandwidth, security, and QoS issues of WLAN. Some of these extensions are reviewed in next chapter.

## 6.6 HomeRF

HomeRF is a wireless protocol architecture designed and developed from scratch to bring wireless networking to residential consumers using RF devices [19, 20]. After an evaluation of users' requirements, HomeRF was designed with the ability to provide simultaneous wireless broadband Internet access, resource sharing, multiple streaming media sessions, and multiple toll-quality voice connections.

HomeRF aims to face the interoperability limitations of many in-home wireless-networking products. It is supported by the HomeRF Working Group (HRFWG), which was formed in early 1997 to establish the mass deployment of interoperable wireless-networking access devices to both local content and the Internet for voice, data, and streaming media in consumer environments. The HomeRF specification defines a new common interface that supports wireless voice and data networking in the home. The standard is based on Proxim's RF LAN technology, which permits transmission speeds of up to 1.6 Mbps.

**Figure 6.22**    NOKIA IEEE 802.11b-GPRS PC card.

HomeRF is based on the Shared Wireless Access Protocol–Cordless Access (SWAP-CA), or SWAP, for radio-based home networks codeveloped by Compaq, HP, Intel, Microsoft, Motorola, Proxim, and Sony. For the design of the SWAP specification, the HRFWG has selected to reuse proven RF networking technology for data and voice communications, simplified for residential usage. In this way, the SWAP specification aims to define a new common air interface that supports both wireless voice via PSTN and VoIP, along with LAN data services and Internet access via TCP/IP in the home environment. SWAP operates in the unlicensed 2.4-GHz ISM band, which is available worldwide; combines elements of the DECT and the IEEE 802.11 standards; supports both a TDMA service to provide delivery of interactive voice and other time-critical services and a CSMA/CA service for delivery of high-speed packet data up to 10 Mbps; and ensures interoperability among various wireless products being developed by PC, communications, and consumer-electronics vendors for the home market.

### 6.6.1  Network Architecture

The HomeRF specification defines four types of SWAP-CA devices:

1. *Connection Point:* The HomeRF base station and operates as an interworking unit (IWU) between SWAP-compatible devices and the wired network (e.g., PSTN, Home LAN, Internet);
2. *Asynchronous Node:* A HomeRF device that communicates using asynchronous traffic (data) (e.g., laptop, PDA, TV set and wireless VCR);
3. *Isochronous Node:* A HomeRF device that communicates using isochronous traffic/streams like voice and video (e.g., cordless phone);
4. *Asynchronous-Isochronous Node:* A HomeRF device that carries both asynchronous and isochronous traffic (e.g., wireless multimedia laptop).

According to the existence or not of a connection point, HomeRF defines two types of network architecture: the *structured architecture* and the *peer-to-peer ad hoc architecture*. The heart of the structured architecture is the connection point, which may be either a standalone device, a peripheral, or an integral part of a home PC or RG. It provides for interworking between asynchronous and/or isochronous SWAP devices and networks like PSTN, HomeLAN or the Internet (Figure 6.21). Isochronous communication is provided via TDMA service, while data communication is supported via CSMA/CA. Moreover, HomeRF may also use the connection point to support power management for longer terminal battery life by scheduling devices' wakeup and polling time.

In the case of peer-to-peer communication, HomeRF supports a network architecture (Figure 6.23) similar to the IEEE 802.11 ad hoc architecture shown in Figure 6.17. It is important to note that the structured architecture supports both data- and

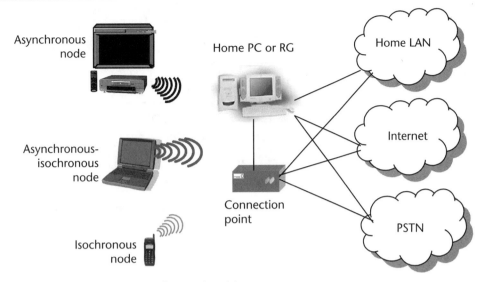

**Figure 6.23** HomeRF structured network architecture.

time-intensive types of traffic, while ad hoc communication supports only asynchronous data communication via CSMA/CA service.

Finally, the SWAP-CA Protocol allows HomeRF users to define multiple overlapping VPNs. In this way, HomeRF devices may be logically organized into independent groups, while being in the same area. The classical application for this feature is the case of block apartments, where the users of each apartment should be able to define their own networks, independent of neighboring apartments. To prevent neighboring networks from interfering with and affecting each other, SWAP-CA defines a 24-bit network identifier (NWID) to distinguish overlapping VPNs. All the devices that belong to the same VPN have the same NWID. The NWID is defined by the home-network administrator, while new devices may learn the NWID when they are connected to the specific HomeRF VPN.

### 6.6.2 Software Architecture

Like most networking interface standards, the HomeRF specification describes the lower two layers of the OSI protocol stack model (Figure 6.24). The Physical Layer has been designed to cover the cost, data rate, and range requirements, while the data link control (DLC) provides data classification and prioritization, security, roaming, and mapping to upper layers. Moreover, both layers are optimized to provide better interference protection and higher network density.

As Figure 6.24 shows, the protocol architecture is close to the IEEE 802.11 WLAN standard in the Physical Layer and extends the DLC Layer with the addition

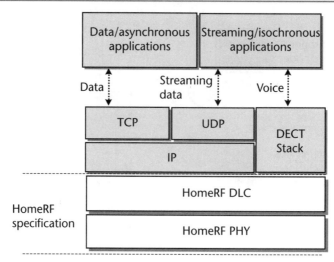

**Figure 6.24**   HomeRF protocol stack.

of DECT to support isochronous services such as voice. As a result, the SWAP DLC Layer can support both data-oriented services, such as TCP/IP, and the DECT/GAP Protocols for voice.

### 6.6.2.1   HomeRF Physical Layer

The HomeRF Physical Layer has been based on IEEE 802.11 FHSS with modifications to reduce costs. SWAP operates at the unlicensed 2.4-GHz ISM band and uses two-level frequency shift key (2-FSK) and four-level frequency shift key (4-FSK) modulations for 0.8-Mbps and 1.6-Mbps data rates, respectively. The low transmitter power and the relaxed receiver sensitivity define a range of up to 50m, which may be reduced to 20m in low-power mode.

### 6.6.2.2   HomeRF DLC Layer

To support both data and voice communications efficiently, the HomeRF DLC Layer combines the IEEE 802.11 MAC Layer with the ETSI DECT standard and supports data transmission and real-time voice communication via the CSMA/CA and TDMA services, respectively.

The structure of the SWAP frame structure is shown in Figure 6.25. Time is organized in superframes with fixed durations, which are decided by the HomeRF devices using a synchronization mechanism.

Just before a superframe transmission starts, all HomeRF devices hop to a new frequency according to a common frequency-hopping pattern. A superframe's transmission starts with a connection point beacon (CPB) time slot, which provides for

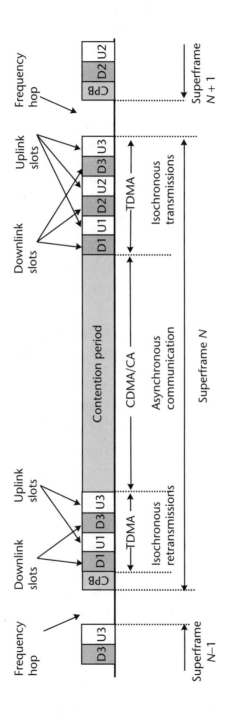

**Figure 6.25**  SWAP frame structure.

network synchronization, superframe control, power management, and the like. Then, a TDMA contention-free period may follow. During this period, retransmission of lost or corrupted isochronous data from previous superframes is transmitted. Time slots are organized in pairs containing one slot for downlink (connection point to device) and one for uplink (device to connection point) communication. In the HomeRF specification, voice is segmented in 20 msec, sampled at 8 KHz, and then compressed and encoded to four-bit adaptive differential pulse code modulation (ADPCM) code words (32 Kbps). For each time slot, a single voice segment is transmitted (20 msec) leading to 640 bits, which is equivalent to a DECT A-field or B-field with some additional bits of control and address data.

Next, a CDMA/CA contention period follows, which is used for the transmission of asynchronous data. The protocol is based on IEEE 802.11, and an acknowledgment is required for every packet. Finally, another TDMA contention-free period transmits voice data. Up to eight time-slot pairs may be sequentially transmitted, but if fewer are present, the extra reservations will be canceled and the time slots will be used for asynchronous transmission of data.

### 6.6.3  HomeRF Security

The frequency-hopping pattern and the VPN NWID provide HomeRF with a security mechanism adequate for the majority of applications. However, neighbors that receive the HomeRF signals may listen to all frequencies and determine the hop-frequency pattern and the NWID. To increase security, SWAP uses the ETSI DECT security model for voice communications, which allows both encryption and roaming between DECT base stations. For data encryption, SWAP provides an optional shared-key encryption algorithm, which provides for security and privacy. The algorithm uses a 56-bit key and a 32-bit initialization vector, which includes part of the MAC address, to encrypt data.

The HomeRF 2.0 standard includes support for 128-bit encryption, so all the data traveling across the radio waves is scrambled.

Large manufacturers (e.g., Motorola, IBM, Intel, Compaq, LSI Logic) have already developed products; however, HomeRF has not yet captured the expected market share, mainly because of its limited bandwidth capabilities. However, the SWAP group has extended the specification to meet the 10-Mbps goal in the 2.4-GHz band and is working towards a definition for SWAP Multimedia (SWAP-MM) that will increase the speed to 20 Mbps in the 5-GHz band.

## 6.7  HIPERLAN

High-Performance Radio Local Area Network (HIPERLAN) is the European reply to the IEEE 802.11 standard [21, 22]. The aim of HIPERLAN is to build a WLAN standard with performance comparable to Ethernet, which would be able to support inherently isochronous channels and guarantee QoS. In 1991, ETSI organized a committee to specify the new high-performance wireless protocol. The major difference from IEEE 802.11 was that HIPERLAN started from scratch as a new specification with specific objectives and was not based on existing products or regulations.

The first draft of the HIPERLAN standard was published in 1995. The standard allowed bandwidth of 23.529 Mbps with support of multihop routing, both asynchronous and time-bounded communication, prearranged or ad hoc network topology, and power saving.

### 6.7.1 Network Architecture

Just like IEEE 802.11 and HomeRF, HIPERLAN supports both a structured network architecture that incorporates a base-station and ad hoc peer-to-peer networking. Moreover, HIPERLAN supports optional packet forwarding. In this case, a node may operate as an intermediate reflector, or *forwarder,* that forwards packets to other nodes. The forwarder node may multicast (broadcast) packets to all HIPERLAN nodes or unicast packets to a specific node. To achieve the latter, the reflector node has to maintain and dynamically update a database, where routing and addressing information is stored. In the absence of sufficient routing information, a unicast packet transfer may also be distributed to all the nodes in the HIPERLAN.

### 6.7.2 Software Architecture

Just like most WLAN standards, HIPERLAN specifies only the two lower layers of the OSI protocol stack, while the upper layers remain unchanged for interoperability reasons (Figure 6.26).

The requirements for high bit-rate and low-power transmission, for reasons of safety and autonomy, limited the standard to distances of up to 100m, while the bandwidth and the multichannel transmission required a large spectrum of 150 MHz. Thus, the Conference of European Posts and Telecommunications Administration (CEPT) selected the 5-GHz band, which was divided into five channels. The lower three channels are available throughout Europe, while the upper two channels are available only in some countries.

The HIPERLAN Physical Layer uses Gaussian minimum shift key (GMSK) and frequency shift key (FSK) technologies to deliver up to 23-Mbps data rates. The MAC Layer is based on a carrier sensing mechanism, but differs from the IEEE 802.11 standard as it introduces packets prioritization via a nonpreemptive priority multiple access (NPMA) mechanism. The HIPERLAN channel access control sublayer supports time-bounded communication through the following algorithm.

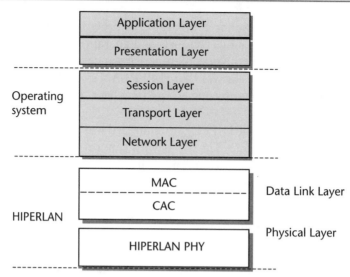

**Figure 6.26**   HIPERLAN protocol stack.

In case the terminal senses that the air interface is idle for a sufficient interval, immediate transmission is allowed. If it is busy, channel access is granted in three phases: the prioritization, elimination, and transmission phases. The prioritization phase excludes nodes with lower-priority packets from further channel access contention. It consists of one to five slots, one per priority category, and each slot is 256 bits long. Following a descending-priority-per-slot strategy, each node transmits a data burst if it has not sensed a data burst of higher priority. During the elimination phase, which lasts for 12 slots or less, each node that had transmitted a burst during the prioritization phase transmits a burst for a geographically distributed number of slots; then, it listens to the channel for one more slot. If it hears another burst, then it stops contending, as a different node has a longer burst to transmit. The contention resolution scheme ensures that each surviving data transmission attempt has a statistically equal chance of gaining transmission permission. Thus, after the elimination phase, only the node(s) with the higher priority and longer burst will still contend for the channel. Finally, in the transmission phase, each of the surviving nodes defers transmission for a geographically distributed number of slots, while listening to the channel. If it senses that the air interface is busy, it completely defers transmission.

### 6.7.3   HIPERLAN QoS

QoS is one of the stronger features of HIPERLAN. QoS is supported via two mechanisms: the *priority* of the packet (high or normal) and the packet *lifetime*. Packet lifetime is measured in milliseconds and the valid range is 0 to 16,000 ms (the default is

500 ms). The priority of the packet, together with its residual lifetime, determines the channel access priority of the packet. The channel access priority falls into one of five categories, which are used during the prioritization phase. Due to the multi-hop packet forwarding mechanism, the lifetime and the residual lifetime of a packet are transmitted along with the packet. If a packet can not be delivered within the residual lifetime it is discarded.

## 6.8  Summary

Table 6.3 summarizes the characteristics of wireless in-home-networking technologies.

As Table 6.3 shows, IrDA and Bluetooth are short-range technologies (up to 10m), DECT and HomeRF are medium-range technologies, and 802.11 and HIPERLAN have ranges greater than 100m. However, IrDA and Bluetooth do not

**Table 6.3**  Summary of Home Wireless Technologies

|  | *IrDA* | *DECT* | *Bluetooth* | *802.11* | *802.11b* | *HomeRF* | *HIPERLAN* |
|---|---|---|---|---|---|---|---|
| *Frequency Band* | IR | 1,880–1,900 MHz (1,900–1,930 MHz) | 2.4 GHz | 2.4 GHz | 2.4 GHz | 2.4 GHz | 5 GHz |
| *Physical* | Optical | Frequency hopping | FHSS | FHSS, DSSS | DSSS | FHSS | GMSK/FSK |
| *Access* | Polling | TDMA | Master-slave, polling | CSMA/CA | CSMA/CA | CSMA/CA + TDMA | NPMA |
| *Maximum Range* | 10 cm–10m | 50m | 10 cm–10m | 150m | 150m | 50m | 100m |
| *Power* | Very low | Low | Very low | Medium | Medium | Medium | Medium |
| *Complexity* | 1x | 1x | 1.5x | 2x | 2x | 3x | 3x |
| *QoS* | No | Yes | Yes | Not inherited | Yes | Yes | |
| *Security* | No | Medium/strong | Medium | Medium | Strong | Strong | |
| *Throughput* | | | | | | | |
| *Physical* | 2.4 Kbps (4 Mbps) | 640 bps (2 Mbps) | 1 Mbps (10 Mbps) | 2 Mbps | 11 Mbps | 1.6 Mbps (10 Mbps) | 23.5 Mbps |
| *Effective* | ≤96 Kbps | ≤552 Kbps | ≤0.7 Mbps | ≤1.5 Mbps | ≤7 Mbps | ≤1.2 Mbps | ≤20 Mbps |
| *Regional Support* | Worldwide | Europe (worldwide) | Worldwide | Worldwide | Worldwide | United States (worldwide) | Europe |
| *Promoters* | 5,000+ | 5,000+ | 2,000+ | 100+ | 100+ | <50 | <20 |
| *Price* | Very low | Low | Low | Medium | Medium | Medium | High |
| *Target Application* | Control and data | Wireless voice | Serial cable replacement | Wireless data | Wireless data | Wireless voice/data | Wireless voice/data |

compete with each other as IrDA has already captured a very large segment of the market for consumer-electronics control applications, while Bluetooth targets new applications in the telecommunications market. Also DECT has captured a very large percentage of in-home telephony services; thus, it is not likely to be replaced in the near future. IEEE 802.11b has replaced IEEE 802.11 as it provides up to 11-Mbps capacity and is backwards compatible with 802.11. IEEE 802.11b is a mature technology, already used in the business sector, but lacks QoS to support voice applications.

Chapter 7 will present the evolution of these technologies and try to sketch the future of wireless home networks.

# References

[1]  Millar, I., "Infrared Standards for High-Speed Infrared Communications," *Hewlett Packard Journal*, Feb. 1998, pp. 1–20.

[2]  IrDA, "Serial Infrared (SIR) Physical Layer Link Specification," Ver. 1.0, The Infrared Data Association, Apr. 27, 1994.

[3]  IrDA, "Serial Infrared Link Access Protocol (IrLAP)," Ver. 1.1, The Infrared Data Association, June 16, 1996.

[4]  IrDA, "Link Management Protocol (IrLMP)," Ver. 1.0, The Infrared Data Association, Aug. 12, 1994.

[5]  IrDA, "Tiny TP: A Flow-Control Mechanism for Use with IrLMP," Ver. 1.0, The Infrared Data Association, Oct. 25, 1995.

[6]  IrDA, "IrCOMM: Serial and Parallel Port Emulation over IR (Wire Replacement)," Ver. 1.0, The Infrared Data Association, Nov. 7, 1995.

[7]  See http://www.dectweb.com.

[8]  ETSI, "Digital Enhanced Cordless Telecommunications (DECT): A High Level Guide to the DECT Standardization," ETSI TR 101 178, Mar. 2000.

[9]  Bluetooth Core Specification at http://www.bluetooth.com.

[10]  Bisdikian, C., "An Overview of Bluetooth Wireless Technology," *IEEE Communications Magazine*, Dec..2001, pp. 86–94.

[11]  Miller, B., and Bisdikian, C., *Bluetooth Revealed: The Insider's Guide to an Open Specification for Global Wireless Communications,* Upper Saddle River, NJ: Prentice Hall, 2001.

[12]  Shepherd, R., "Bluetooth Wireless Technology in the Home," *IEE Electronics and Communications Engineering Journal*, Oct. 2001, pp. 195–203.

[13]  Zahariadis, T., Pramataris, K., and Zeros, N., "A Comparison of Competing Broadband In-Home Technologies," *IEE Electronics and Communications Engineering Journal*, Aug. 2002, pp. 195–203.

[14]  Zyren, J., and Petrick, A., "Brief Tutorial on IEEE 802.11 Wireless LANs," Intersil, Feb. 1999, at http://www.intersil.com/an9829.pdf.

[15]  3Com, "IEEE 802.11b Wireless LANS," technical paper, at http://www.3com.com.

[16]  ISO/IEC, "IEEE 802.11 Local and Metropolitan Area Networks: Wireless LAN Medium Access Control (MAC) and Physical (PHY) Specifications," ISO/IEC 8802-11:1999(E).

[17]   Brewer, B., et al., "Security of the WEP Algorithm," at http://www.isaac.cs.ber-
       keley.edu/isaac/wep-faq.html.

[18]   Grimm, B., "Overview of Wi-Fi Protected Access," at http//:www.wi-fi.org/opensec-
       tion/pdf/Wi-Fi_protected_access_overview.pdf.

[19]   HomeRF, "Wireless Networking Choices for the Broadband Internet Home," white paper,
       at http://www.homerf.org.

[20]   HomeRF, "Quality of Service in the Home Networking Model," white paper, at
       http://www.homerf.org.

[21]   LaMaire, R., et al., "Wireless LANs and Mobile Networking: Standards and Future Direc-
       tions," *IEEE Personal Communications,* Aug. 1996, pp. 86–94.

[22]   ETSI TC-RES, "Radio Equipment and Systems (RES; High Performance Radio Local-Area
       Network (HIPERLAN); Functional Specification," ETSI 06921, Sophia Antipolis
       Cedex–France, Draft ETS 300 652, July 1995.

# Emerging Wireless Technologies

## 7.1 Introduction

In the previous chapter, we reviewed a number of existing wireless in-home-networking technologies. However, user and application requirements for more bandwidth, isochronous channels, guaranteed QoS, greater coverage, and advanced security have spurred the evolution of existing technologies or the specification of emerging new standards.

One evolutionary technique that really changed the bandwidth capabilities of wireless transmission technologies is OFDM [1]. OFDM is a modulation technique that uses multiple carriers separated at precise frequencies to mitigate multipath effects (i.e., the reception of the same signal on more than one path). In this way, OFDM splits the channel bandwidth into a large number of subchannels such that the channel frequency response is essentially flat over the individual subchannels. By using OFDM-coded carriers, the wireless system avoids intercarrier interference (ICI), as subcarriers are orthogonal to each other. Moreover, the system may be more resistant to channel impairments (i.e., multipath fading or narrowband interference) as the coded data is spread across all the carriers; therefore, if some carriers are lost, the information can be reconstructed from the error-correction bits in other carriers [2].

In this chapter, we are going to describe the evolution of some important existing wireless technologies and some new technologies that are expected to play a significant role in the future of home networks.

## 7.2 Evolution of IrDA

For years, IR technologies have been available for in-home control and data communications. IrDA provided a standard common protocol for interoperability between devices from different manufacturers. The low cost and flexibility of the IrDA standard has motivated many manufacturers to propose evolutionary paths towards increased range and lower cost. Among them are IrDA very fast infrared (VFIR), IrDA Lite, and IrOBEX.

IrDA VFIR is an enhancement of IrPHY, which extends IrDA bandwidth capabilities [3]. The initial version of IrDA included only the IrPHY 1.0 serial infrared (SIR) specification, which defines a serial communication link for distances of 1m to 3m and data rates between 2,400 bps and 115.2 Kbps. The bandwidth of the SIR link is enough for control applications, but too narrow for data-LAN applications. Aiming to solve the bandwidth problem, the IrPHY 1.1 standard enables data rates of up to 4 Mbps by specifying additionally to SIR, fast infrared (FIR) channels.

Due to further competition with RF standards and between IR manufactures, HP, IBM, and Sharp jointly proposed an extension of IrDA with the addition of a new channel, called a *VFIR channel* [4]. The IrDA VFIR was approved by the IrDA association in March 1999 and supports data rates of up to 16 Mbps, which is four times the speed of IrDA FIR and comparable to the bandwidth provided by modern RF technologies. IrDA VFIR achieves the same maximum link distance and field of view as previous versions of IrDA. Moreover, it does not require any special or expensive optical components. The VFIR specification has been demonstrated using commercially available light emitting diodes (LEDs), while the price for the 16-Mbps full implementation is expected to be comparable to the Bluetooth implementation (on the order of $5).

Another extension of IrDA is IrDA Lite, which targets low-cost IR devices and describes the design methodology for a minimal IrDA-compliant device implementation, aiming to further lower their cost. To achieve this, the specification simplifies the IrLAP and IrLMP modules, eliminates all optional features [i.e., sniffing, role exchange, universal interface (UI) frames], supports many predefined values (e.g., 9,600-bps data rate, 64-byte data size, 500-ms turnaround time), ignores frames that do not have the correct address, uses simplified algorithms for discovery, and the like.

As a result, IrDA Lite specifies a system with minimal code, data, and complexity, resulting in minimal read only memory (ROM), random access memory (RAM), and processing power, respectively. However, IrDA Lite is a recommendation rather than a standard. An IrDA Lite implementation does not need to follow all IrDA strategies, but it may lie anywhere between a minimal IrDA Lite to a full-featured IrDA device.

Another important IrDA extension is IrOBEX, which provides a generic method for application object sharing and exchange between different IrDA devices. In this way, IrOBEX simplifies application development, as all low-level IrDA communication functions (e.g., link discovery, establishment, maintenance) are transparent to the programmer, who can focus on the application's specific requirements.

IrOBEX is based on HTTP, but is more compact. Along with the IrDA communications and the object exchange, it piggybacks object description headers. The peer IrDA device can use the additional information and handle the objects intelligently.

## 7.3  Evolution of Bluetooth

Bluetooth 1.0 was intended to serve as a universal, low-cost, wireless cable replacement. Thus, throughput, coverage range, interoperability, and security were not among the major requirements. However, during the maturing phase of the standard, new requirements appeared.

The first was interoperability between Bluetooth-compliant devices from different vendors. For security reasons, Bluetooth devices communicate via encrypted links. During the establishment of a new link, the devices exchange public keys to confirm their identities [5]. The Bluetooth 1.0 specification describes the operation, but leaves important details open to the manufacturer's interpretation. As a result, Bluetooth 1.0–compliant devices from different manufacturers may generate different encryption keys and fail to negotiate the initial link establishment. The problem appears if each device believes it is the master communication device. Bluetooth 1.1 solves this initial interoperability problem by requiring the slave devices to acknowledge the master device and confirm their role as slaves.

Another problem was related to the protocol frame structure. Bluetooth 1.0 supports optionally up to five slots per packet to achieve the maximum data-transfer rate of 720 Kbps per channel, but not all Bluetooth devices support the five-slot format option. In a Bluetooth communication, if the master transmits more slots per packet than the slave can support, the communication will fail. However, the Bluetooth 1.0 specification does not foresee a control protocol with which slave devices can inform the master device of their communications capabilities and of the maximum number of slots they can support. In Bluetooth 1.1, the slave is able to acknowledge the master, inform it about packet sizes, and control the communications flow.

Finally, Bluetooth 1.0 was under pressure to guarantee worldwide operation and compliance with all countries' frequency plans. Bluetooth is an FHSS technology, which defines a 79-hops pattern in the 2.4-GHz frequency band. However, the frequency plans of some countries, including Japan, France, and Spain, use the 2.4-GHz frequency for noncommercial purposes (e.g., military communications). To extend coverage in these countries, Bluetooth 1.0 specified a different hopping pattern with 23 hops, which does not use the specific frequencies of the 2.4-GHz spectrum. However, Bluetooth devices that use the 79-hops pattern are incompatible with those that follow the 23-hops pattern; thus, they cannot be used in these countries. To overcome this interoperability issue, BSIG managed to gain worldwide permission for the 79-hop-pattern equipment. In this way, the 23-hop option was removed, and Bluetooth 1.1–compliant devices use only the 79-hops pattern to communicate within the 2.4-GHz frequency.

The market requirement for greater coverage range and broader communications is evolving Bluetooth 1.1 into Bluetooth 2.0. Bluetooth 2.0 is expected to be capable of 20-Mbps transfer rates across ranges of up to 50m. Moreover, the direct integration of Bluetooth 2.0 chips into mobile terminals (e.g., PDAs and phones) will offer users the ability to use interchangeably local Bluetooth connections, wherever available, or to roam to the third generation (3G) and 2.5G mobile networks. This will enable greater flexibility, making personal access networks (PAN) widely available.

Bluetooth has captured a small percentage of the personal communications market, primarily because Bluetooth chips are in most cases embedded in other devices. The flexibility, simplicity, and low cost of the Bluetooth networking solution, as well as of the extensions to and new versions of the protocol, are expected to increase the volume of Bluetooth products rapidly within a few years. Based on a study from TDK Systems Europe, Ltd. [6], the number of Bluetooth-enabled products is expected to grow to 1.8 billion units by 2006 (Figure 7.1), while another study from Cahners In-Stat Group gives even better results, projecting 1.3 billion units by 2005.

## 7.4   Evolution of IEEE 802.11

IEEE 802.11 has for years been the most mature solution for WLAN communications; thus, the general term *Wireless LAN* or *WLAN* refers in most cases to the IEEE 802.11 standard or its evolution, IEEE 802.11b. Initially, IEEE 802.11 was operating at 2.4 GHz, providing data rates of up to 2 Mbps without any inherited QoS. Soon the IEEE 802.11b Physical Layer specification was defined, which achieves data rates of up to 11 Mbps.

The wide acceptance of IEEE 802.11/802.11b, the competition, and user demand for broader, more secure, and guaranteed QoS communications necessitated IEEE to create a number of task groups aimed to enhance the initial standard. Each task group has specific objectives related to the protocol's bandwidth, frequency band, security, QoS, and interoperability.

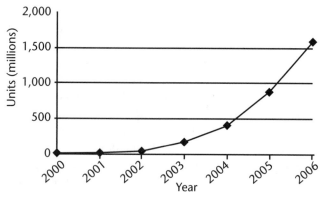

**Figure 7.1**   Estimation of installed Bluetooth products [6].

The more mature evolutions are the IEEE 802.11a and the 802.11g. Since early 2001, the IEEE 802.11g task group has formed a draft standard that achieves data rates higher than 22 Mbps, adopting either single-carrier trellis-coded eight–phase shift keying (PSK) modulation or OFDM schemes. The 802.11a technology operates in the unlicensed 5.8-GHz band and supports data rates of up to 54 Mbps using OFDM schemes. In parallel, other 802.11 task groups aim to enhance specific areas of the protocol. The following IEEE 802.11 task groups are currently active:

- *802.11a* provides 54-Mbps capability, uses the 5.8-GHz spectrum range, and has been on the market since 2002.

- *802.11d* works towards 802.11b versions at other frequencies for countries where the 2.4-GHz band is not available. Thanks to the ITU's recommendation and extensive lobbying by equipment manufacturers, most countries have already released this band.

- *802.11e* works towards the specification of a new 802.11 MAC Protocol to accommodate additional QoS provision and security requirements over legacy 802.11 Physical Layers. It replaces the Ethernet-like MAC Layer with a coordinated TDMA scheme and adds extra error-correction to important traffic. A key part of 802.11e is the 802.1p standard, which provides a method to differentiate traffic streams in priority classes to support streaming applications.

- *802.11f* aims to improve the handover mechanism in 802.11 so that users can maintain a connection while roaming between access points attached to different networks.

- *802.11g* aims to define a new standard, which will use the 2.4-GHz frequency band, feature full compatibility with already deployed 802.11b devices, and provide date rates of up to 22 Mbps. IEEE 802.11g is expected to go to market in 2003.

- *802.11h* aims to enhance control over transmission power and radio channel selection of the 802.11a standard, to be acceptable by the European regulators.

- *802.11i* aims to enhance 802.11 security. Instead of WEP, a new authentication/encryption algorithm based on the Advanced Encryption Standard (AES) will be proposed.

- *802.11j* will propose the issue of 802.11a and High Performance Radio Local Area Network Type 2 (HIPERLAN2) interworking.

To ensure interoperability and compatibility across all market segments, IEEE 802.11 product manufactures have agreed on a compliance procedure called *Wi-Fi*. Moreover, WECA has been formed to certify Wi-Fi interoperability of new products.

As Figure 7.2 shows, the number of installed IEEE 802.11 products is expected to increase rapidly over the next few years.

## 7.5  IEEE 802.11a

IEEE 802.11a is one of the most promising evolutions of IEEE 802.11b; thus, we study it in more detail. The 802.11a standard is similar to 802.11b, but provides wireless data speeds of up to 54 Mbps and uses the 5.8-GHz spectrum range, which experiences less interference than the 2.4-GHz spectrum.

### 7.5.1  Physical Layer

The Physical Layer of IEEE 802.11a is a multicarrier system based on OFDM. It uses 52 carriers; of these, 48 are data carriers, which carry user data traffic, and 4 are pilot carriers, which are used for synchronization and system control purposes. Each carrier is 300 KHz wide. The total bandwidth occupied by the 52 carriers along with the channel spacing is 20 MHz [7].

IEEE 802.11a uses various modulation schemes, namely BPSK, QPSK, 16 QAM, and 64 QAM with 1/2 or 3/4 error-correcting code overhead. According to the modulation, each of the 48 data carriers may transmit raw data rates from 125 Kbps to 1.5 Mbps, so the total raw bandwidth may vary from 6 Mbps to 72 Mbps. Assuming that 64 QAM is used for maximum bandwidth (72 Mbps), reduced by 3/4 error-correction code overhead, IEEE 802.11a may achieve up to 54-Mbps useful traffic.

### 7.5.2  Medium Access Control Layer

IEEE 802.11a uses a MAC Protocol almost identical to IEEE 802.11b that is based on CSMA/CA. An IEEE 802.11a terminal must initially sense the medium for a

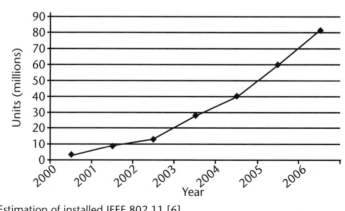

**Figure 7.2**  Estimation of installed IEEE 802.11 [6].

specific time interval, and if the medium is idle, it can start transmitting the packet. If the medium is not idle, the terminal begins a back-off process and waits for another time interval (a minimum of 34 $\mu$sec). When the back-off time has expired, the terminal can try to access the medium again. As collisions in wireless environments cannot be detected, a positive acknowledgement is used to notify that a frame has been successfully received.

Figure 7.3 shows the format of a packet-data unit in 802.11a, including the preamble, header, and Physical Layer service data unit (PSDU, or payload). The preamble is mainly used for frame synchronization. The header is mapped on 1 OFDM symbol and is coded with the most robust modulation and error-correction scheme (BPSK 1/2). It contains information about the type of modulation and the coding rate used in the rest of the packet (Rate field), the length of the payload (Length field), and the parity bit. The Tail field is used to reset the encoder and provide information to the decoder. The Service field is currently used to initialize the descrambler (seven bits), while nine bits are reserved for future use. The Pad field is used for mapping the physical payload data unit (PPDU) to an integer number of OFDM symbols.

IEEE 802.11a products are already available. However, there are certain barriers before IEEE 802.11a gains worldwide acceptance. First of all, the coverage range is very short. The distance of 50m is just adequate for in-home usage. Another reason is the selection of the 5-GHz frequency band. This band is not available worldwide. Japan, for example, permits the use of a smaller band, containing half the channels. In Europe, the standard does not comply with various European Union (EU) requirements. Moreover, IEEE 802.11a does not provide any QoS mechanisms. Thus, Europe is promoting the HIPERLAN2 standard, as it guarantees QoS.

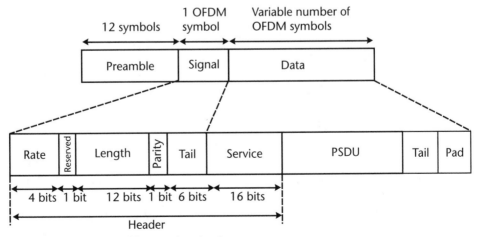

**Figure 7.3**   IEEE 802.11a frame and packet format.

All these constraints still restrain the large-scale adoption of IEEE 802.11a, limit the volume of production, and keep costs high. A step in the direction of wide establishment of IEEE 802.11a is the creation of a multivendor interoperability certification for 802.11a products. Since November 29, 2002, WECA has provided a Wi-Fi equivalent certification for IEEE 802.11a products.

## 7.6   IEEE 802.15.3

The IEEE 802.15.3 is a wireless specification designed from scratch to support ad hoc networking and multimedia QoS guarantees. It was designed to achieve data rates from 11 to 55 Mbps, targeting distribution of high-definition video and high-fidelity audio. IEEE 802.15.3 is specified by the IEEE 802.15.3 High Rate (HR) Task Group (TG3) for wireless personal area networks (WPANs). The group aims to specify a new standard for high-rate (20 Mbps or greater) WPANs. The new standard will provide for high data rate, low power, and low cost solutions, addressing the need of portable mulitmedia applications.

IEEE 802.15.3 focuses on a low-range ad hoc network topology. The distance between the devices should be short, typically less than 10m. Communication between IEEE 802.15.3 devices is centralized and connection-oriented. During the initialization phase, one device of the WPAN is selected to be the master or coordinator of the network. The coordinator is the center of the network, and its role is to provide network synchronization, admission control, and management of the piconet network resources and power save.

### 7.6.1   Protocol Stack

The IEEE 802.15.3 protocol stack is shown in Figure 7.4. Like most wireless network technologies, IEEE 802.15.3 specifies only the Physical and the Medium Access Layers. To facilitate smooth integration with existing network protocols, 802.15.3 specifies a variety of service-specific convergence sublayers (SSCS) such as IP, IEEE 1394, and USB. Moreover, it provides hooks for the specification of SSCS for other emerging protocols.

The IEEE 802.15.13 Physical Layer operates in the unlicensed frequency band of 2.4 GHz. The selection of the 2.4-GHz band is highly important, since the 5-GHz band is prohibited for outdoor use in many countries, including Japan. The IEEE 802.15.13 Physical Layer occupies the 15-MHz bandwidth, which allows for up to four fixed channels and an 11-Msps symbol rate. According to channel conditions, application requirements, and device capabilities, IEEE 802.15.3 may use five different types of modulation formats. The base modulation format is Trellis-coded QPSK

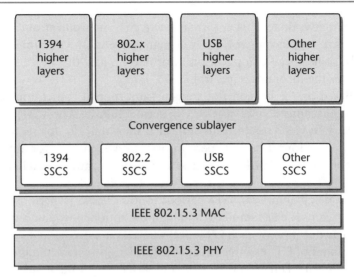

**Figure 7.4**   IEEE 802.15.3 protocol stack.

at 11 Mbps. Other alternatives are uncoded QPSK at 22 Mbps and 16/32/64-QAM with eight-state two-dimensional trellis coding at 33, 44, and 55 Mbps, respectively. Since early 2001, the IEEE 802.15.3a Study Group (SG3a) has worked towards the provision of a higher-speed Physical Layer enhancement amendment to the 802.15.3 standard, one capable of reaching speeds as high as 480 Mbps for shorter ranges.

The IEEE 802.15.3 Physical Layer is optimized for short ranges, enabling low costs and integration into small consumer devices (e.g., a flash card or a PC card). The Physical Layer also requires low current drain (less than 80 mA) while actively transmitting or receiving data and minimal current drain in the power-save mode.

The IEEE 802.15.3 MAC is based on a TDD/TDMA scheme. Time is organized in slot, and slots are grouped in superframes. The IEEE 802.15.3 superframe is shown in Figure 7.5. Initially, a network beacon is transmitted. The beacon slot

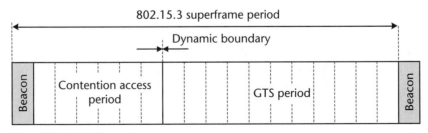

**Figure 7.5**   IEEE 802.15.3 superframe structure.

carries network-specific information (e.g., synchronization, information for new devices to join the network, power management, and maximum transmitted power level), some specific application information, and the allocation of dedicated time slots for the incoming superframe.

Based on the beacon slot transmission time, a contention access period (CAP) and a guaranteed time slot (GTS) period follow. The CAP is optional and uses CSMA/CD MAC for transmission of frames that do not require guaranteed QoS. CAP is used for all management information exchange, for authentication/association and channel time (CT) request and response, and for exchange of asynchronous data.

Finally, a number of GTS periods follow. The GTS periods are used for isochronous streams such as standard and high-definition video, high-quality audio, and the like. The duration of each GTS slot is based on a predefined set of QoS policies and the amount of CT resources available for data transfers and is specified in the initial beacon frame. During a GTS slot, a terminal may send an arbitrary number of data frames as long as the aggregate duration of the transmissions does not exceed the scheduled duration.

### 7.6.2 QoS Support

IEEE 802.15.3 achieves QoS using a per-flow reservation scheme. All signaling and QoS negotiation is left to the service- or network-specific higher layers. IEEE 802.15.3 just provides enough information, so Layer 3 or the upper layer is able to request and negotiate the right level of resources. Each device has to communicate with the network coordinator to discover if a specific QoS requirement can be granted. In the case of positive response, the coordinator allocates an appropriate amount of GTS CT to this node and assigns a flow identifier for that specific flow between the IEEE 802.15.3 SSCS and the IP Layer.

By delegating QoS negotiation to higher layers and making resource reservation, IEEE 802.15.3 provides very good resource planning and allocation, supports various signaling protocols, and reduces significantly the complexity of the lower layers. However, in some cases, peer-signaling communication may increase the negotiation overhead and result in large delay.

### 7.6.3 WiMedia Alliance

The IEEE 802.15.3 is supported and promoted by the WiMedia Alliance, which was launched in September 2002 as a not-for-profit open industry forum, aiming to develop and adopt "standards-based specifications for connecting personal area, wireless multimedia devices." The companies in the alliance are involved in various industry sectors, including chips, imaging, document management, PCs, and wireless networking. The founding members include Appairent Technologies, Kodak,

HP, Motorola, Philips, Samsung Electronics, Sharp, Time Domain, and Xtreme-Spectrum. At the time of this writing, more vendors are expected to join the group in the coming months. The WiMedia Alliance's target is to create wireless specifications based on the MAC Layer of IEEE 802.15.3.

IEEE 802.15.3 specifies a Physical Layer using 2.4-GHz spread-spectrum technology, but WiMedia members are hoping to combine the MAC Layer with an Ultrawideband Physical Layer for a superfast wireless multimedia standard (Figure 7.6). They have also proposed a new Transport and Control Layer, which will support synchronized PtMP distribution of multimedia streams in addition to asynchronous data. Finally, an SDP will provide for flexible connections and auto-configuration of the devices.

## 7.7  HIPERLAN2

HIPERLAN2 is the European proposal for a broadband WLAN operating with data rates of up to 54 Mbps in the Physical Layer on the 5-GHz frequency band. HIPERLAN2 is supported by ETSI and was developed by the Broadband Radio Access Networks (BRAN) group [8].

HIPERLAN2 is a flexible radio LAN standard designed to provide high-speed access to a variety of networks, including 3G mobile core networks, ATM networks, and IP-based networks, and also for private use as a WLAN system. HIPERLAN2 is a connection-oriented TDM Protocol. Data is transmitted on connections that have been established prior to the transmission using the signaling functions of the HIPERLAN2 control plane. This makes implementing support for

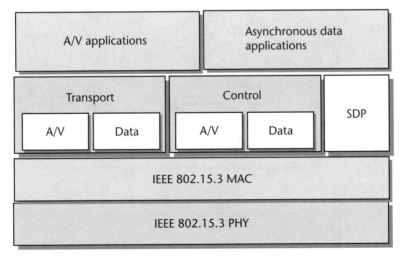

**Figure 7.6**  WiMedia protocol stack.

QoS straightforward. Each connection can be assigned a specific QoS, for instance in terms of bandwidth, delay, jitter, BER, and the like. It is also possible to use a more simplistic approach, where each connection can be assigned a priority level relative to other connections. This QoS support in combination with the high transmission rate facilitates the simultaneous transmission of many different types of data streams, (e.g., video, voice, and data).

### 7.7.1  Software Architecture

Figure 7.7 shows the HIPERLAN2 protocol stack. At the Physical Layer HIPERLAN2 uses OFDM to transmit the analog signals in the 20-MHz bandwidth [9]. Just like IEEE 802.11a and IEEE 802.15.3, it uses BPSK, QPSK, 16-QAM and 64-QAM modulation and 1/2, 2/3, and 3/4 error-correcting overhead. Moreover it uses 52 OFDM carriers, 48 for data and 4 as pilots. In this way it may achieve 12-, 24-, 48-, or 72-Mbps coded bit rates and data rates of up to 54 Mbps.

The difference between the HIPERLAN2 and the IEEE protocols lies in the DLC Layer [10]. In HIPERLAN2, the DLC contains the MAC Layer, EC and DLC control modules, and convergence sublayers. The MAC Protocol is built from scratch, implementing a type of dynamic TDMA/TDD scheme with centralized control.

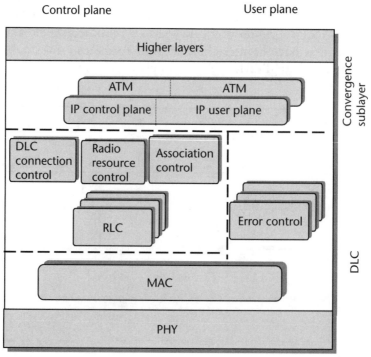

**Figure 7.7**  HIPERLAN2 protocol stack.

Figure 7.8 shows the basic MAC frame structure. It has a period of 2 msec, which is dynamically divided into a downlink period (from the central station to the terminal stations) and an uplink period (from the terminal stations to the central station). The durations of the uplink and downlink periods depend on the traffic to be sent. The downlink periods starts with the broadcast channel, or BCH, which contains control information (e.g., transmission power levels, wakeup indicators, HIPERLAN2 network and access-point identifiers) that is broadcasted to all mobile terminals. Then the frame control channel (FCH) follows, which describes the structure of the frame [i.e., how resources are allocated, when the downlink and uplink periods begin, and where the random access channel (RCH) is located]. The access feedback channel (ACH) carries information on previous access attempts made in the RCH.

Next, the downlink and uplink data periods begin; these are the main transmission periods. Each time slot may contain a long DLC user PDU (54 bytes long with a payload of 48 bytes), or a short control PDU (9 bytes long). The last part of the frame is the RCH. The RCH is used by the terminals to request transmission resources for the next MAC frame. If collisions occur during RCH access, they are reported back to the terminals in the next ACH.

HIPERLAN2 provides a selective repeat ARQ EC mechanism. The EC detects errors in the user PDUs and requests retransmission of packets that have been received in error or lost. Packets are acknowledged positively or negatively and delivered in sequence to the Convergence Layers. To support QoS for delay-critical applications, such as voice, the EC may discard any data packets that become obsolete and all PDUs with lower sequence numbers that have not been acknowledged.

The DLC Control Plane provides DLC management and control functionality. It contains the radio link control (RLC), which provides a transport service to the DLC user connection control, the radio resource control (RCC), which controls the radio resources, and the association control function (ACF), which allocates identifiers to the mobile terminals.

Finally, various convergence sublayers are specified (e.g., Ethernet, IP, ATM, UMTS) to facilitate smooth integration with the existing network. The convergence

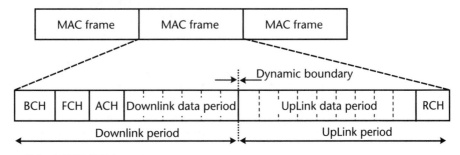

**Figure 7.8**   HIPERLAN2 frame structure.

sublayers provide the core network with a common RLC sublayer. Moreover, they provide all functions for connection setup, translation of QoS requirements to RRC parameters, mapping of network PDUs from one network protocol to one RLC entity, and removal of redundant network PDU control information (header compression and decompression).

Figure 7.9 shows the structure of the long, or user, PDU; it is 54 bytes long and contains, first, control information (1.5 bytes or 12 bits), then the PDU Type field (2 bits) and the PDU sequence number (10 bits). The packet is protected with cyclic redundancy check (CRC-24), which is 3 bytes long. The remaining 49.5 bytes are fielded with the payload [11].

HIPERLAN2 is a well-designed specification, which promises adequate bandwidth and QoS. However, it is a European standard that has not yet gained popularity and wide acceptance outside European borders. Implementations of the protocol have been announced and products are expected in early 2003.

## 7.8   5-GHz Unified Protocol

IEEE 802.11a technology is one of the most powerful and mature wireless home-networking technologies. However, it does not provide any inherited QoS support; thus, it is not approved by the European regulators who prefer ETSI's HIPERLAN2. To overcome this issue, ETSI and IEEE have formed the 5-GHz Partnership Project (5GPP), a joint venture that aims to merge 802.11a and HIPERLAN2 into a single standard known as the 5-GHz Unified Protocol (5-UP) [12]. By tying two or even three channels together, this standard would offer even higher data rates than existing systems. Three channels will provide a real throughput of about 100 Mbps, more than most laptop PCs can handle, while securing European approval for a future version of 802.11a.

The 5-UP proposal extends the OFDM system to support multiple data rates and usage models. 5-UP will enhance the existing IEEE 802.11a standard by permitting cost-effective designs in which everything from cordless phones to high-definition televisions and PCs could communicate on a single wireless multimedia network. 5-UP will achieve this by allocating the carriers within the OFDM signal on an individualized basis. As Figure 7.10 shows, multiple devices simultaneously

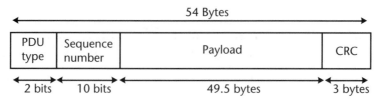

**Figure 7.9**   Long PDU format.

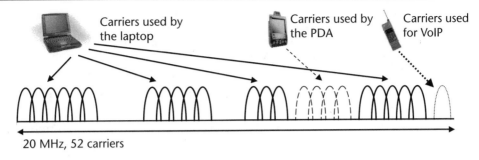

Figure 7.10   Example of 5-UP scalable carrier allocation.

transmit to an access point using different OFDM carriers. By zeroing out some inputs to an inverse fast Fourier transform, some carriers may be handled by other devices.

## 7.9   Summary

Table 7.1 provides a summary of the emerging in-home wireless technologies. IrDA and Bluetooth target short-range communications, while the other technologies target higher-bandwidth and longer-range applications. Among them, IEEE 802.11a and HIPERLAN2 are directly competitive technologies. Although HIPERLAN2 is superior as far as QoS and complexity are concerned, IEEE 802.11a is better promoted and 802.11a-compliant products are already available.

**Table 7.1**   Comparison of Emerging In-Home Wireless

|  | IrDA (VFIR) | Bluetooth 2 | 802.11a | 802.11g | 802.15.3 | HIPERLAN2 | 5-UP |
|---|---|---|---|---|---|---|---|
| Frequency Band | IR | 2.4 GHz | 5 GHz | 2.4 GHz | 2.4 GHz | 5 GHz | 5 GHz |
| Data Rate | 16 Mbps | ≤10 Mbps | ≤54 Mbps | ≤22 Mbps | ≤55 Mbps | ≤54 Mbps | ≤100 Mbps |
| Range | Line of sight | 10 cm–10m | Up to 50m with bandwidth degradation | | 10m ad hoc networking | ~20m | ~10m |
| Current Drain | <10 mA | <30 mA | <350 mA | < 350 mA | < 80 mA | N/A | N/A |
| Complexity | 1x | 1.2x | 4x | ~3.5x | 1.5x | 1.5x | 5x |
| QoS | Guaranteed | Guaranteed | Inherited in 802.11e; backwards compatibility is questionable | | Guaranteed | Guaranteed | Guaranteed |
| Video Channels | 1–2 | 1 | 10 | 2–3 | 10 | 10 | 20 |

The main advantage of IEEE 802.11g and IEEE 802.15.3 is that they operate in the unlicensed 2.4-GHz frequency band. This is highly important, since the 5-GHz band is prohibited for outdoor use in many countries, including Japan. However, 802.15.3 focuses on ad hoc wireless networking, while 5-UP is a very promising technology that aims to merge IEEE 802.11a and HIPERLAN2, but that is quite complicated.

As a result, it is not clear which technology will finally capture the largest in-home market segment. IEEE 802.11a and 5-UP have the political advantage; however, it is my opinion that wireless technologies will finally dominate in-home networking.

# References

[1] Sari, H., and Karam, G., "Orthogonal Frequency-Division Multiple Access and Its Application to CATV Networks," *European Transactions on Telecommunications (ETT)*, Vol. 9, No. 6, Nov.–Dec. 1998, pp. 507–516.

[2] Balech J.-P., and Sari, H., "Advanced Modulation Techniques for Broadband Wireless Access Systems," *Proc. Seventh European Conference on Fixed Radio Systems and Networks (ECRR 2000)*, Dresden, Germany, Sept. 2000, pp. 159–164.

[3] van Nee, R., et al., "New High-Rate Wireless LAN Standards," *IEEE Communications Magazine*, Vol. 37, No. 2, Dec. 1999, pp. 82–88.

[4] Bisdikian, C., "An Overview of Bluetooth Wireless Technology," *IEEE Communications Magazine*, Dec. 2001, pp. 86–94.

[5] Phillips, M., "Reducing the Cost of Bluetooth Systems," *IEE Electronics and Communications Engineering Journal*, Oct. 2001, pp. 204–208.

[6] Hunn, N., "Bluetooth and Wi-Fi: The Market Status," TDK Systems Europe, Ltd, March 2002, at http://www.newswireless.net/articles/020326-phoney.html.

[7] ISO, "Local and Metropolitan Area Networks—Specific Requirements—Part 11: Wireless LAN Medium Access Control (MAC) and Physical Layer (PHY) Specifications, High Speed Physical Layer in the 5 GHz Band," ISO/IEC 8802-11:1999/Amd 1:2000(E), 2000.

[8] Johnsson, M., "HIPERLAN2—The Broadband Radio Transmission Technology Operating in the 5 GHz Frequency Band," HIPERLAN2 Global Forum, 1999.

[9] ETSI, "Broadband Radio Access Networks (BRAN); HIPERLAN Type 2 Functional Specification; Data Link Control (DLC) Layer; Part 1: Basic Data Transport Function," ETSI Report TR 101 761-1, Ver. 1.1.1, April 2000.

[10] ETSI, "Broadband Radio Access Networks (BRAN); HIPERLAN Type 2; Data Link Control (DLC) Layer; Part 2: Radio Link Control (RLC) Sublayer," ETSI Report TR 101 761-2, Ver. 1.1.1, April 2000.

[11] ETSI, "Broadband Radio Access Networks (BRAN); HIPERLAN Type 2; Physical (PHY) Layer," ETSI Report TR 101 475, Ver. 1.1.1, April 2000.

[12] McFarland, B., et al., "The 5-UP Protocol for Unified Multiservice Wireless Networks," *IEEE Communications Magazine*, Vol. 39, No. 11, Nov. 2001, pp. 74–80.

# CHAPTER 8

# Technologies with New Wiring Requirements

## 8.1 Introduction

One of the most daunting costs of home networking is installing new cabling. Putting data cables into a new building is an expensive, and increases the overall cost, but not prohibitively so. On the other hand, rewiring an existing home or apartment is difficult and is not a viable solution for the mass market, especially for brick or stone houses. Moreover, most consumers are unwilling to pay for or cannot afford such a large-scale home rewiring.

However, the networking solutions with special wiring requirements provide a secure way to deploy new services in home-networking environments, because they have been tested beforehand for years in the enterprise and business sectors. Therefore, it is expected that a large percentage of home-networking solutions will be based on technologies that have requirements for new home wiring. These solutions will not form the home backbone network, but will extend the home distribution network with cord-patches, in-home restricted SOHO LANs, and peripherals interconnections [1]. Among various interfaces and technologies that fall into this category, we emphasize the Ethernet and USB technologies.

## 8.2 Switched Ethernet

IEEE 802.3 is the well-known Ethernet standard. Ethernet is the default wired-LAN technology for PCs and may be the most widespread LAN technology used in corporate and enterprise networks. The wide acceptance of Ethernet is expected to make it one of the major candidate technologies for in-home LANs.

### 8.2.1 Ethernet LAN Topologies

Initially, Ethernet started as a bus topology over Bayonet Neill Concellman (BNC) coaxial cable for corporate environments with speeds of up to 10 Mbps (Figure 8.1).

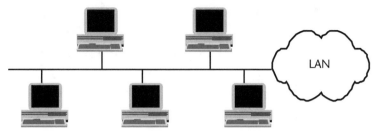

**Figure 8.1**   Ethernet bus topology.

Terminals were tapped on the cable, without any specific plan, creating performance and fault-isolation problems.

As Ethernet LANs continued to grow, the flat bus topology changed to a more structured star topology. Nodes are organized in shared (multistation) or dedicated (single-station) segments, and segments are interconnected with bridges, routers, or hubs. The star topology (Figure 8.2) increases installation costs, as it limits the number of ports per hub, but it helps with fault isolation and provides a more structured methodology for expanding LANs. Moreover, higher speeds can be achieved as the number of collisions on the cables is dramatically lower. To further increase bandwidth and isolate faults, user PCs migrate toward single-station segments.

Bottlenecks occur in star topologies mainly at the interconnection of segments. Routers provide a reliable solution, but the cost is prohibitive for single-station segments. A low-cost solution for node interconnection using a star Ethernet topology is provided by LAN switches, which can boost bandwidth on overburdened, traditional LANs, while working with conventional cabling and adapters, because they allow for simultaneous switching of packets between ports.

### 8.2.2   Switching Technologies

There are three primary types of switching technologies: store-and-forward, switch-direct, and combined. *Cut-through switching* begins forwarding packets as

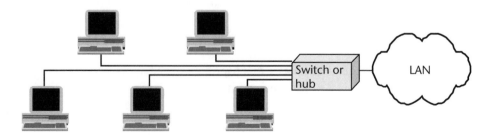

**Figure 8.2**   Ethernet star topology.

soon as the switch receives the first bytes of a frame and reads the packet destination. *Store-and-forward switching* buffers incoming bytes until the full packet is received. Then, it performs CRC to verify the integrity of the packet, and if no errors are found, the packet is forwarded to the destination port.

Cut-through switching reduces transmission latency between ports, but it may broadcast in bursts to the destination port. On the other hand, store-and-forward-based switching reduces erroneous packets and avoids collisions that may effect the overall performance of the segment, but it adds latency to normal processing times.

Some switches perform on both levels. They begin with cut-through switching and monitor the number of errors with CRC checks. After a certain threshold has been passed, they become store-and-forward switches until the number of errors declines. This type of switching is called *threshold detection* or *adaptive switching*.

Finally, some state-of-the-art switches use a new technology called *tag switching*. Tag switches apply an additional short tag in the packet's header and forward the packets between switch ports based on their tags instead of their headers. The tags are matched by hardware using a Tag Information Base lookup table, which achieves very fast Layer 2 switching instead of Layer 3 routing.

### 8.2.3   Fast Ethernet Standards

Although 10-Mbps switched Ethernet should be adequate for most SOHO applications, home users can create fast Ethernet LANs that run at 100 Mbps with minimum additional cost. Fast Ethernet keeps the format and MAC Protocol of the 10-Mbps Ethernet, but increases the transmission speed.

In June 1995, the Fast Ethernet or IEEE 802.3 100BASE-T standard was approved. It covers three standards: 100BASE-TX, 100BASE-T4, and 100BASE-FX.

- *100BASE-TX* uses two pairs of Category 5 UTP cable, providing full-duplex operation at distances of up to 100m. 100BASE-TX is the most widespread Fast Ethernet standard in corporate environments, and it is also expected to be widely used for SOHO applications.

- *100BASE-T4* uses four pairs of Category 3, 4, or 5 UTP cable, providing half-duplex operation at distances of up to 100m. Three twisted pairs are used for data transmission, and one is reserved for collision detection. Currently, 100BASE-T4 is not very popular. However, its evolution, 100BASE-T2, will operate over two pairs of Category 3 UTP cable and may replace 100BASE-TX especially for home usage.

- *100BASE-FX* operates over multimode, two-strand fiber at distances of up to 2 km. 100 BASE-FX may not be used for in-home networks as fiber is not a very flexible medium, but it may be use for home access.

Other Ethernet standards include the following:

- *100VG-AnyLAN* uses four pairs of Category 3 and 5 UTP cable, providing full-duplex operation. The 100VG-AnyLAN specification employs a completely new MAC Protocol to support Ethernet and Token Ring frame formats and offers not only data, but voice and video transmission as well ("VG" stands for "voice-grade").
- *100BASE-SX* is a low-cost solution, which operates over rather short-distance fiber (up to 300m) and provides a migration path towards Gigabit Ethernet.

Ethernet is going to be one of the major alternatives for in-home networks. It is a mature technology with simple installation and configuration, supported by a plethora of vendors, which have driven the cost of the devices, NICs, and wiring way down. It is no coincidence that the majority of STBs and cable and DSL modems have an Ethernet interface or that many telecom operators request home-networking products with multiple-switched 10/100 BASE-T Ethernet interfaces.

Figure 8.3 shows a selection of different Ethernet cards that have PCI interfaces. The wide acceptance of Ethernet and the requirement to minimize costs have motivated vendors to provide more flexible solutions. For example, Figure 8.4 shows a 10/100BASE-T PC card and an integrated Ethernet and modem PC card. Moreover, Figure 8.4 shows a PCI card that operates as an Ethernet interface and a Fast Ethernet Switch, eradicating the cost of an external additional switch.

Of course, Ethernet is not a home-networking panacea. It has several drawbacks, starting with the need to rewire the home with Category 5 UTP cable and the need for a NIC for each PC. Moreover, home networks will require two types of data transfer: *asynchronous* and *isochronous*. Asynchronous transport is used for traditional computer file transfers. A request for data is sent to a specific address and an acknowledgment is returned. Multimedia applications such as video, audio, and voice telephony, however, require QoS. If there are delays in the data, the receiving device must buffer the signals, creating cost and resynchronization problems.

Ethernet does not inherently support QoS for isochronous streams. Thus, inherently it cannot support multimedia applications like video or voice distribution. Two approaches can be taken to solve this problem: using QoS techniques like

    10 BASE-Tx       10 BASE-2     10/100 BASE-T   Dual 100 BASE-Fx

**Figure 8.3**   Different types of Ethernet PCI cards.

10/100 BASE-T PCMCIA
with external RJ-45

10/100 BASE-T PCMCIA
and 56K modem

4x 10/100 BASE-T
PCI card

**Figure 8.4** Advanced Ethernet cards.

DiffServ, IntServ, and MPLS to control the available bandwidth resources or using switched Ethernet network devices that isolate the traffic between different subnetworks. In this way, large files may be downloaded or transferred in one segment without harming the quality of a video stream in a different segment.

Another potential problem with in-home video-distribution applications over Ethernet has to do with the intellectual property rights of the content. Ethernet is a broadcasting medium network. To control distribution, content providers have formed a Copy Protection Technical Working Group (CPTWG), which tries to lock movies using content encryption. In this way, although all users receive the content, only subscribers are able to watch it.

Regardless of these drawbacks, Ethernet is projected to play an important role in home networking, especially for new homes.

## 8.3 USB

The USB [2] standard is an external extension bus that simplifies interconnection of peripherals. USB aims to replace all the different kinds of serial and parallel port connectors with one standardized PnP interface. The main advantage of USB in home networks is its ability to "hot-swap" multiple peripherals in "daisy-chain" architecture. As most PCs have at least two USB ports onboard, accessible outside the case, there is no need to open up the computer to connect a new peripheral. Moreover, most operating systems provide device drivers for USB in their standard distributions; thus, connecting new USB devices is a rather automatic process. The USB cable connectors are shown in Figure 8.5.

### 8.3.1 USB Features

The following advantages of USB can be identified [3]:

- *PnP/Hot-Swappable Peripheral Connectivity:* In USB, "connecting a new peripheral is as simple as plugging a new telephone in a telephone jack" [2].

**Figure 8.5**   USB cable connectors.

USB connections do not require terminator settings, ID numbers, memory addresses, or interrupt settings. From the software point of view, only a device driver is required, which is preinstalled on most operating systems.

* *Simple Network Architecture:* The USB network topology is very simple (Figure 8.6). The network is based on a PC with at least two USB ports. Devices are connected directly to a USB port on the PC or to another USB device forming a chain. If all ports are filled up, a USB hub may provide additional ports (usually 4 or 7). USB supports up to 127 devices simultaneously by attaching peripherals through interconnected external hubs.

* *Higher Performance:* USB 1.1 supports data-transfer rates of up to 12 Mbps. Increased bandwidth is translated into a larger set of more flexible peripherals and increased productivity. For example, USB desktop conferencing cameras can provide larger pictures and smoother motion; scanners can scan faster;

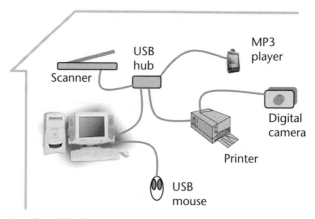

**Figure 8.6**   USB network architecture.

and mass-storage devices like an external hard disk or CD writer can be connected using a USB link.

• *Asynchronous and Isochronous Data Transfer:* Home networks will require both ATM for data and isochronous transfer mode for A/V streams, voice-telephony, and the like. USB 1.1 provides both asynchronous data and isochronous streaming channels for compressed video distribution.

• *Power Supply:* It is said that there are never enough ac power outlets in any computing environment. USB addresses this concern by providing an in-line power supply to small devices through the USB data cable. It enables automatic sensing of power requirements and delivers the requisite power to the devices, often eliminating the need for additional power supplies. For example, devices such as cameras can be connected through a single cable and receive both power and data signals.

USB was initially specified in 1995. Politically, it is supported by many very large PC hardware manufacturers, software vendors, and the USB Implementers Forum, which are trying to establish USB as the de facto standard for all peripheral devices. It is no coincidence that all PCs made before 1997 have no USB ports, but that all PCs made after 1998 almost certainly support at least two USB ports. In fact, USB has become a key enabler of the Easy PC Initiative, an industrywide initiative led by Intel and Microsoft to make PCs easier to use. This effort sprung out of the recognition that users need simpler, easier-to-use PCs that do not sacrifice connectivity or expandability.

Moreover, very large manufacturers like Intel, Microsoft, Compaq, HP, Lucent, NEC, and Philips are among the companies that support USB and led the development of the USB 2.0 specification [4]. USB 2.0 is backwards compatible with USB 1.1, which uses the same cables, connectors, and software interfaces, but provides transfer rates of up to 460 to 480 Mbps, which is about 40 times faster than USB 1.1 and enables bandwidth-demanding devices, such as digital video-conferencing cameras, next generation scanners and printers, and DVD drives.

USB is enjoying tremendous success in the marketplace, with most peripheral vendors around the globe developing products to this specification (Figure 8.7). USB will also play an important role in home networks, mainly for short-distance connections between PCs and peripherals.

### 8.3.2 A Comparison of USB and Ethernet

USB is a superior technology as compared with Ethernet. It provides inherited QoS, higher bit rates, and in-line power supply. However, Ethernet is more mature, and its wide acceptance in business segments is expected to turn it into one of the major candidate technologies for in-home LANs. Last but not least, USB targets different

| | USB 2.0 external | USB 2.0 | USB 2.0 |
| USB 2.0 hub | hard disk drive | web camera | Flash disk |

**Figure 8.7**   A small selection of USB products.

applications and very short distances, while Ethernet can be part of the in-home backbone network.

## References

[1]   Zahariadis, T., Pramataris, K., and Zervos, N., "A Comparison of Competing Broadband In-Home Technologies," *IEE Electronics and Communications Engineering Journal (ECEJ),* August 2002, pp. 133–142.

[2]   See USB at http://www.usb.org.

[3]   Compaq, Intel, Microsoft, NEC, "Universal Serial Bus Specification," Rev. 1.1, Sept. 1998.

[4]   USB, "An Introduction to Hi-Speed USB," at http://www.usb.org/developers/usb20/developers/whitepapers/usb_20g.pdf.

# CHAPTER 9

# Firewire

## 9.1 Introduction

This chapter focuses on IEEE 1394, or Firewire, an emerging, low-cost, digital in-home bus. IEEE 1394 aims to integrate entertainment, communication, and computing electronics into consumer multimedia networks.

IEEE 1394 requires the installation of new wires in the home. However, it incorporates a number of advanced features that cannot be underestimated. First of all, IEEE 1394 devices may communicate directly, without a supervising PC, at speeds of up to 400 Mbps; a roadmap has been drawn to increase the bandwidth to 3.2 Gbps. Another important feature is that IEEE 1394 is a completely digital technology. It does not convert data to an analog signal, but transmits it in digital format. Thus, it can improve and maintain the signal integrity. Other IEEE 1394 features include PnP and Hot Swap capability, isochronous data channels, and power-supply provisioning through the same data cable [1].

## 9.2 IEEE 1394

IEEE 1394 is a new serial interface for broadband in-home communications. It was initially proposed for entertainment appliances, but soon turned into an emerging standard that targets all consumer multimedia networks.

The technology was first introduced by Apple Computers as Firewire in 1986. Later, Sony referred to it as i.LINK and trademarked that name. In the meantime, the IEEE adopted the standard as IEEE 1394. Currently, it is promoted by the 1394 Trade Association, formed in September 1994 to accelerate the market adoption of IEEE 1394. As of this writing, the association has more than 170 member companies, including representatives from the consumer-electronics, computer, and semiconductor industries. The steering committee of the 1394 Trade Association consists of Adaptec, AMD, Apple Computer, Cirrus Logic, Compaq, IBM, Microsoft, Molex, Maxtor, Mitsubishi, NEC, National Semiconductor, Philips, Sony, Sun, and Texas Instruments [2].

Initially, in 1995, the IEEE 1394 standard supported data rates of up to 50 Mbps. In 2000, IEEE 1394a was approved. It revises and extends IEEE 1394 and supports speeds of 100, 200, and 400 Mbps. IEEE 1394a also specifies a cable interface and traffic control and adds power-management features. The evolution of 1394a, IEEE 1394b is already underway and aims to provide for data-transfer rates of up to 3.2 Gbps and full backwards compatibility with the current specification.

IEEE 1394 implements an easy-to-set-up, hot-plugging, scalable, bus architecture. Furthermore, it supports both asynchronous and isochronous data transfer and guarantees data transport at a predetermined rate using isochronous data channels.

### 9.2.1  Cable and Topology

The IEEE 1394 standard specifies two IEEE 1394 cables: one with four copper wires and one with six (Figure 9.1). Both cables have two twisted-wire pair cables that carry the signals. Each twisted pair is shielded, as it is the entire cable. The difference between the two is that the six-wire cable includes a power-supply pair. This pair may carry 8V to 40V dc at up to 1.5 Amps, which allows the maintenance of the device's Physical Layer continuity when the device is powered down or during malfunctions, or the powering other devices in the network via a bus topology. The four-wire cable lacks this powering cable pair; thus, four-pin devices cannot be powered via the IEEE 1394 cable, but must have an external power source.

On the other hand, the dimensions of the six-pin connector are $10 \times 5$ mm, almost double that of the four-pin connector, which measures only $5 \times 3$ mm (Figure 9.2). Thus, the four-pin connector is better suited for portable devices. The maximum cable length for applications with bandwidth requirements greater than 200 Mbps is only 4.5m. However, at slower speeds, cables lengths up to 14m can be used, while there are efforts to extend this distance to 25m [3]. IEEE 1394b may support data rates of up to 3.2 Gbps over distances of 100m using optical fiber.

Although IEEE 1394b does not specify a physical medium, in most cases it uses copper plastic optical fiber (POF). POF provides the advantages of glass optical fiber (GOF), but at lower cost and with easier use. Initially, GOF received more attention,

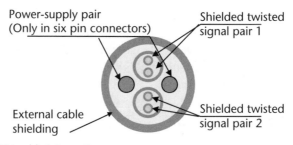

**Figure 9.1**   IEEE 1394 cable intersection.

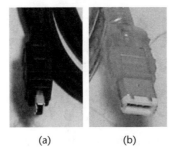

(a)                    (b)

**Figure 9.2**   IEEE 1394 cable connectors: (a) four pins, and (b) six pins.

due to its rapid acceptance in the telephone industry, but recent developments have made POF a contender for a large number of data and audio applications.

Up to 63 IEEE 1394 devices may be connected in a star or tree configuration. Multiple devices can be daisy-chained off of any branch. Figure 9.3 shows an example IEEE 1394 network topology. The standard does not permit more than 16 cables or 72m distance between any nodes in the network topology. However, this number maybe even larger with the use of repeaters to extend the communication distance. IEEE 1394 bridges may interconnect more than 1,000 bus segments, thereby providing large growth potential.

These limitations are derived from the address structure of the nodes. Each node is identified by a 16-bit ID, containing a 10-bit bus ID, which allows up to 1,024 bus segments to be identified, and a six-bit node, or physical ID, which allows up to 64 nodes to be attached on the same bus. As Figure 9.4 shows, the ID has an additional 48 bits for addressing a specific offset within each IEEE 1394 node's memory. This allows direct copy of data onto a specific memory location of the destination node. The 64-bit addressing scheme also conforms to the Control and Status Register (CSR) bus architecture standard.

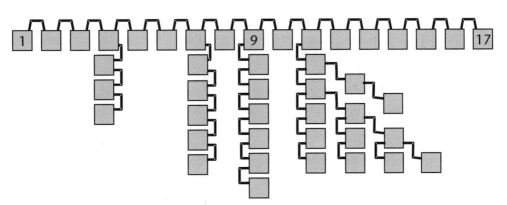

**Figure 9.3**   Example IEEE 1394 network topology.

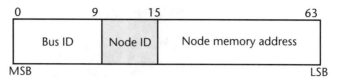

**Figure 9.4**  IEEE 1394 address structure.

IEEE 1394 features a PnP capability, which eliminates the need for address switches or other user intervention when reconfiguring the bus. An IEEE 1394 device may be added to or removed from the bus at any time, even when the bus is in full operation. Addressing is done dynamically each time the system is powered up, or when a device is added or removed from the network. There is no cable terminator required or addressing ID to be set, while the electrical contact is made inside the connectors, thus preventing any potential electrical shock to the user.

### 9.2.2   Protocol Stack

IEEE 1394 specifies the lower protocol stack layers, starting from the physical medium interface up to the Network Layer. Figure 9.5 shows the IEEE 1394 protocol stack.

As it is shown, the IEEE 1394 specifies three layers: the Physical Layer, the Link Layer, and the Transaction Layer. The Physical Layer is always implemented in hardware. It resides over the physical medium interface and connects the device to the actual medium connector. It is responsible for all Physical Layer functions (e.g., bus initialization, encoding/decoding, arbitration). There are two Physical Layer specifications: Physical Layer A provides 100, 200, or 400 Mbps, while Physical Layer B will provide up to 3.2 Gbps. It is important to note that the IEEE 1394 protocol stack is independent of the actual Physical Interface.

The IEEE 1394 Link Layer is located just above the Physical Layer and is also implemented in hardware. It corresponds to the ISO/OSI MAC Layer and is responsible for packet transmission and reception of both isochronous channels and asynchronous packets, and for EC using CRC. Finally the Transaction Layer corresponds to the ISO/OSI LLC Layer and is responsible for packet read, write, and lock functions. Isochronous packets do not pass through the Transaction Layer, but are forwarded directly between the Link Layer and the node upper layers to minimize latency and guarantee timing requirements.

In parallel, the serial bus management plane controls and manages bus functions and isochronous resources. The Bus Management and Transaction Layers are implemented in hardware or firmware.

IEEE 1394 can support both isochronous and asynchronous traffic over the same medium in the same frame. The IEEE 1394 frame/cycle structure is shown in Figure 9.5. The duration of each cycle is 125 $\mu$sec and starts with a cycle start packet

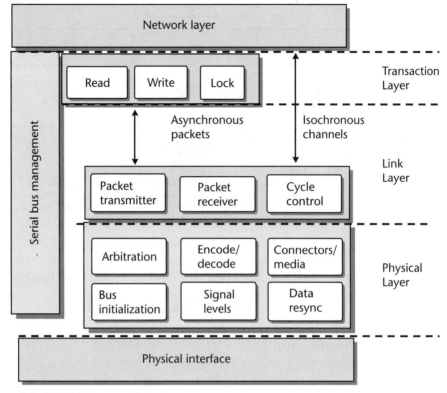

**Figure 9.5** IEEE 1394 protocol stack.

that indicates the start of the new cycle. Isochronous traffic has priority over asynchronous traffic. Thus, after the start-timing indicator, the isochronous period starts. This period is allocated to time slots, and each time slot carries the traffic of a specific channel. For example, in Figure 9.6, time slots 1, 2, 3, 4, and 5 carry the isochronous traffic for channels 2, 4, 5, 8, and 7 respectively.

If the cycle has additional time available, asynchronous traffic is transmitted. In contrast to isochronous time-slot allocation, asynchronous packets and acknowledges are sent to explicit addresses on the 1394 bus.

Figure 9.7 shows a simplified version of the asynchronous write packet format. The packet starts with the Destination ID field, followed by the Source ID field, which contain the 16-bit addresses of the destination and source nodes, respectively. The Transaction Type field describes the packet type (i.e., it is an asynchronous write). The Destination Offset field contains the remaining 48 bits of the destination address as required for CSR addressing. The Transaction Label field is like a sequence number, which allows requests to be matched with acknowledgements. The Data Length field shows the actual length of the data, as the payload may be up to a maximum length according to the transmission speed. For instance, in the

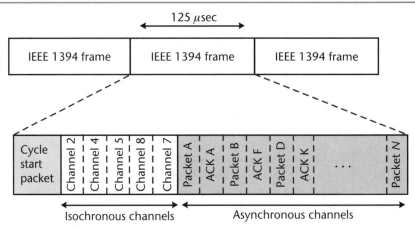

**Figure 9.6**   IEEE 1394 frame structure.

200-Mbps IEEE 1394 version, the payload may be up to 1,024 bytes long, in steps of four bytes. To achieve this, a Pad field is used. Finally, both the header and the data parts of the packet are protected by CRC for error checking.

## 9.3   IEEE 1394 Application

Figure 9.8 shows a typical IEEE 1394 operation. A digital video camera sends a video stream to a digital TV set. The TV stores the video in a VCR and forwards it to a PC, which selectively captures frames for additional processing. The camera is powered directly by the IEEE 1394 cable, while control information is carried as asynchronous packets. The video stream is carried in digital format as isochronous traffic. Each device can receive and process the video directly as digital data, and there is no need for a video capture card or analog-to-digital video conversion. IEEE 1394 guarantees just-in-time delivery of data without the need for collision detection or another arbitration scheme.

While in operation, a Firewire printer is connected to the VCR to make hard copies of some video frames. The PnP mechanism of IEEE 1394 is initiated, a new ID is dynamically assigned, and the device is recognized by the rest of the IEEE 1394 nodes.

## 9.4   Comparison and Interoperability

IEEE 1394 is quite flexible. It multiplexes many asynchronous and isochronous digital signals and provides user-friendly capabilities like PnP, autoconfiguration, and Hot Swap. As a result, it is compared with many in-home-networking alternatives. The major drawback of IEEE 1394 is that it requires new wires, but this is not a big problem in the case of small-distance connections. IEEE 1394, like USB and

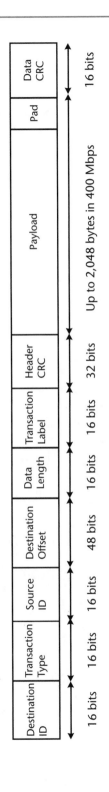

**Figure 9.7** IEEE 1394 asynchronous write packet format.

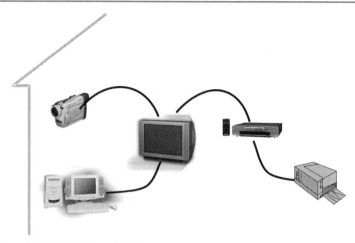

**Figure 9.8**  A typical IEEE 1394 application.

Ethernet, may not target the home backbone network, but rather, extensions of the home distribution network in the form of cord-patches, in-home restricted SOHO LANs, and interconnected peripherals.

As Table 9.1 shows, IEEE 1394b's characteristics are superior to most competitive technologies. Gigabit Ethernet and USB 2.0 provide comparable bandwidth capabilities. However, the former does not support inherited QoS, which is required for isochronous streams and multimedia applications like video and voice distribution, while USB messages have to be processed by a PC, meaning that USB has an additional bottleneck. On the other hand, USB is a cheaper solution that has already been adopted in many PC-oriented products. Thus, it is possible that USB will be preferred for communications where a PC is mandatory, while Firewire will be preferred for multimedia applications.

The popularity of IEEE 1394 has motivated HIPERLAN2 to build a 1394 convergence sublayer (CL) into the HIPERLAN2 standard specification. As 1394 is

**Table 9.1**  Comparison Between IEEE 1394 and Competing Technologies

|  | Actual Bandwidth | Guaranteed Bandwidth | $10^{-12}$ BER | Isochronous QoS | No-New-Wires Requirement | User Friendliness |
|---|---|---|---|---|---|---|
| Powerline | 50 Kbps | No | No | No | Yes | No |
| Phone Line | 10 Mbps | No | No | No | No | No |
| IEEE 802.11b | 10 Mbps | No | No | No | Yes | No |
| 10 Base-T | 6 Mbps | No | No | No | No | No |
| 100 Base-T | 50 Mbps | No | No | No | No | No |
| Gigabit Ethernet | 500 Mbps | No | No | No | No | No |
| USB 2.0 | 480 Mbps | No | Yes | Yes | No | PnP |
| 1394b | 3.2 Gbps | Yes | Yes | Yes | No | PnP |

independent of the Physical Interface, 1394 traffic can be directly transmitted over the air. The restricted data rate of HIPERLAN2, as compared with that of 1394, is managed by the 1394.1 bridge specification, which is located over the HIPERLAN2 DLC, filters the bus traffic, and sends over the air link only the traffic that addresses wireless terminals. Moreover, 1394 CL contains a complete bridging specification, which allows HIPERLAN2 to look like a virtual 1394 terminal, as shown in Figure 9.9.

## 9.5   IEEE 1394 Future

IEEE 1394 has already been adopted as the computer interface of many digital cameras and digital video applications, and it is expected to capture a large percentage of the in-home-networking market, especially in the entertainment domain. Many silicon vendors already provide IEEE 1394 in single-chip solutions. An indicative collection is shown in Figure 9.10. Moreover, many vendors and manufacturers, including Sony, Sun, Philips, Apple, NEC, and Microsoft, have already presented a large variety of products based on Firewire/i.Link.

Many standardization bodies and committees have accepted IEEE 1394 as the standard interface for digital communications. Apart from the 1394 Trade Association [4], Digital VCR Conference (DVC) and the European DVB have endorsed IEEE 1394 as their digital television interface and have proposed IEEE 1394 to the

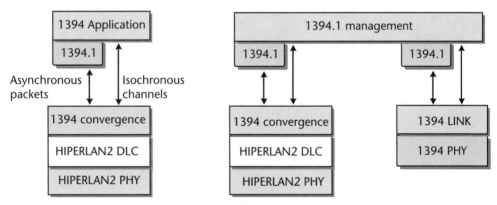

**Figure 9.9**   IEEE 1394 and HIPERLAN2 interoperation.

**Figure 9.10**   IEEE 1394 single-chip solutions.

Video Experts Standards Association (VESA) [5] as the digital home-network medium of choice. The Electronic Industries Association subcommittee, EIA 4.1, has voted for IEEE 1394 as the PtP interface for digital TV and the multipoint interface for entertainment systems. In the United States, ANSI has defined a Serial Bus Protocol (SBP) to encapsulate the small computer systems interface (SCSI-3) for IEEE 1394 [6].

# References

[1]  Hoffman, G., "IEEE 1394. The A/V Digital Interface of Choice," at http://www.1394ta.org/ Technology/About/digital_av.htm.

[2]  Information Gatekeepers, "Plastic Optical Fiber," newsletter, at http://www.igigroup.com/ nl/pages/pof.html.

[3]  Evans, D., "In-Home Wireless Networking; An Entertainment Perspective," *IEE Electronics and Communications Engineering Journal,* Oct. 2001, pp. 213–219.

[4]  IEEE 1394 Trade Association at http://www.1394ta.org.

[5]  Gelman, A., "Challenges of Consumer Communications and Mass Market Information Networking: Closing the Digital Divide," Panasonic Technologies, 2001, at http://www. research.panasonic.com.

[6]  IEEE 1394 Forum at http://www.1394forum.org.

# CHAPTER 10

# Middleware Technologies

## 10.1 Introduction

In the previous chapters, we have already reviewed and analyzed various wireline- and wireless-networking technologies that promise to offer indoor data and voice networks connectivity and access to the broadband superhighway. Due to cost issues, it is expected that in most cases, these technologies will be gradually adopted. The incremental installation and variety of consumer-electronics products will create the future home network as a collection of different physical media and lower-layer technologies. Home PCs, digital STBs, DVD players, and consumer-electronics-network-aware appliances from different manufacturers and following different standards will be connected to isolated home networks. Interoperability between these products will be a very critical issue for the success of the digital home. It will offer the home network the ability to share processing and storage resources, drastically increase control flexibility and monitoring of information exchange, enable multimedia and streaming content management and playback, and introduce new value-added services to home users.

As multiple physical media and lower-layer standards and protocols are expected, interoperability can be provided only at the higher layers of the protocol stack. The software technologies, architectures, and systems that promise seamless interconnection to devices with different hardware and network characteristics are often called *middleware* [1]. Middleware technologies aim to provide a minimum common set of interfaces and functions that will hide devices' or resources' heterogeneity and enable different resource types to communicate transparently. Various home-networking middleware technologies have already evolved. This chapter discusses a selection of standardization efforts and ad hoc approaches by vendors, which, I believe, are among the major players in the field of middleware technology.

## 10.2 Home Electronic System

The Home Electronic System (HES) is a collection of standards that aim to specify rules to guarantee, in a multivendor/multiapplication context, free implementation and configuration of networking devices and interoperation between devices in a home network [2]. The HES working group is part of the International Organization for Standardization (ISO)/International Electrotechnical Commission (IEC)/ Joint Technical Committee 1 (JTC 1), responsible primarily for information standards. The basic principle of HES is to separate devices from network technologies and define stable interfaces between them [3, 4].

The HES standard does not cover the medium and the associated access units. Instead, it gives the specifications for the services delivered through these interfaces and for the local implementations of them. A home network may integrate one or more different physical media (e.g., powerline, twisted-pair, IR, or RF) and may be connected to multiple access networks (e.g., telephone, CATV, power network). An implementation of HES may be assembled by one application at a time, starting from single applications, like lighting control, security control, or audio and video control, to grow eventually into an integrated multiapplication system.

The HES standard aims to provide "plug-compatibility," which is generally understood to mean that the devices can be physically interconnected and will be able to work together. More formally, plug-compatibility incorporates device interconnectivity and interoperability issues. *Interconnectivity* is defined as the ability of the devices to be hooked together on a shared transmission medium, and it is conceived as a sufficient mechanical, electrical, and functional (lower-layer protocol) characteristic of the devices. *Interoperability* is defined as the ability of the higher communication layers of devices connected to different networks to exchange commands, resulting in meaningful actions. The functionality of the devices on both sides of the interface has no direct relationship to the interface as such.

### 10.2.1  HES Interfaces

To make it possible to add and to change existing devices, as well as to enlarge and to upgrade the home network keeping the existing devices, the HES standard defines two types of interface between the home network and the devices: the UI and the process interface (PI). The UI provides full HES functionality, while the PI is meant for simple devices that do not require the full implementation of the HES application protocol. For cheaper devices or network combinations, HES also supports devices that do not have a UI or PI interface, but are connected directly to the medium. These devices will, however, be medium-dependent (conformance type B) and do not have the advantages of devices that conform fully to HES (conformance type A). HES interfaces are shown in Figure 10.1.

By using these interfaces, a manufacturer can design a device both to meet his specific marketing objectives and to provide the option of integrating it into a multiapplication HES, taking advantage of the synergy between different applications. As Figure 10.2 shows, the HES network architecture consists of HES devices with a UI or PI connected to the network via a corresponding network adaptation unit (NAU).

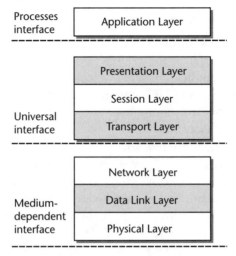

**Figure 10.1**   HES interfaces.

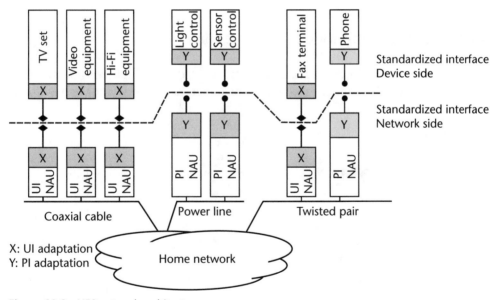

**Figure 10.2**   HES network architecture.

The NAUs will be medium-dependent and provide the HES standard interface on the network side. It is important to note that any device UI may be connected to any network UI, and any device PI to any network PI.

### 10.2.2   HES Application Model

HES assumes that most of its application processes will be distributed. Thus, apart from the HES interfaces, the HES Working Group has specified two fundamental components: the application service elements (ASEs), which allow the user process to communicate via the HES communication system, and a communication language for ASE-to-ASE communication in the same appliance or appliance-to-appliance communication.

Finally, the HES reference model (Figure 10.3) distinguishes communication channels as specifically information or control channels. Control channels use packet-switched transmission, whereas information channels typically use circuit-switched transmission. Packet-switched transmission is fundamental to HES, and all HES implementations provide a packet-switched control channel.

Via the separation between the device/application and the network medium layer with standardized interfaces, HES has introduced great flexibility and portability, as devices may be plugged from one medium and network into another. The HES standard is quite complicated; thus, it has not yet proliferated as expected in the market.

## 10.3   Home Audio-Video Interoperability

Home Audio-Video Interoperability (HAVi) is a higher-layer standard aiming to allow all manner of digital consumer electronics and home appliances to

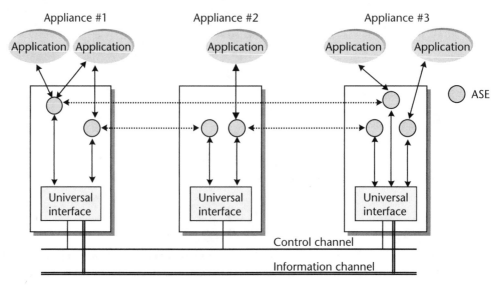

**Figure 10.3**   HES reference architecture.

communicate with each other [5–7]. HAVi has been defined by eight leading manufacturers: Grundig AG, Fraunhofer FOKUS, Hitachi, Panasonic, Philips, Sharp, Sony, Thomson, and Toshiba, who are committed to developing products following the HAVi Protocol. As of 2002, 18 other participants had already joined: Digital Harmony Technologies, Epson, Infineon Technologies, QNX Systems, Loëwe Kenwood, LG Electronics, Mitsubishi Electric, Pioneer, Samsung, Sanyo, Seiko, Sun Microsystems, Tao Group, Teralogic, Vivid Logic, 3A International, Wind River Systems, and Yaskawa Information Systems.

The main objective of HAVi is to provide a lightweight, open, platform-independent, home-networking software specification, for seamless interoperability among home-entertainment products. The exchange of high-quality digital video and high-fidelity audio requires a dedicated network able to provide higher bandwidth and meet strict, real-time constrains. The HAVi specification is A/V device-centric; it has been designed and optimized to meet the particular demands of digital audio and video. It defines an operating-system-neutral middleware that manages multidirectional A/V streams, event schedules, and registries, while providing application-programming interfaces (APIs) for the creation of a new generation of software applications.

The following lists the main characteristics of the HAVi specification [8]:

- *Distributed Control:* Under the HAVi architecture, there is no single master controlling device. Any HAVi device located anywhere within the HAVi network may control other devices, while at the same time being controlled.

- *Interoperability:* The HAVi specification enables entertainment products from different manufacturers to communicate with each other and exchange stream data and control information.

- *PnP:* HAVi defines an autoinstallation and configuration procedure. Periodically or on demand, HAVi-compliant devices announce their presence and capabilities, greatly simplifying the installation and setup procedure.

- *Upgrade Capability:* Within the HAVi specification, new functionalities and capabilities may be downloaded or uploaded to HAVi-compliant entertainment devices via the Internet.

HAVi has selected the IEEE 1394 bus over other physical and transmission protocols to interconnect A/V devices, because it supports multiple isochronous communication channels. The term *isochronous* refers to 1394's ability to guarantee streaming data delivery at fixed intervals. Of course, the future ability to integrate other home networks, such as existing analog links, telephone lines, or new wireless technologies into the HAVi network, was built into the HAVi architecture from the beginning, and bridges to these networks are likely to be developed in the future.

Based on A/V and execution capabilities, HAVi classifies consumer-electronics devices into four categories:

1. *Full A/V Devices:* This category includes devices that incorporate the complete HAVi software suit and are capable of downloading and executing HAVi applications. They act as controllers or execution platforms for other lower-cost devices.

2. *Intermediate A/V Devices:* The HAVi devices included in this category have fewer hardware capabilities, cost less, and can execute limited applications. They do not provide a JAVA runtime environment and cannot act as controllers of the network.

3. *Base A/V Devices:* A base A/V device does not have HAVi middleware software, but hosts a JAVA-based control module instead. When the device is plugged into the network, this module is uploaded and executed from any full A/V device. In most cases, this module will offer read-only information about the base A/V device.

4. *Legacy A/V Devices:* These do not support HAVi, but can be controlled by other devices. These devices use non-HAVi or proprietary protocols and require a full or intermediate A/V device to operate as a gateway between it and the HAVi network.

For example, a modern STB may be a full or intermediate A/V device, a video camera may be a base A/V device using the STB as an execution environment, and a VCR may be a legacy A/V device operating via the STB.

### 10.3.1  HAVi Software Architecture

The HAVi specification defines the software architecture shown in Figure 10.4. HAVi is independent of the physical medium and lower-layer protocols and operates over vendor-specific real-time operating systems (RTOS). However, a specific communication media manager (CMM) per physical medium is defined.

CMMs for protocols that provide for isochronous communication channels, like Bluetooth and USB, are defined. However, as IEEE 1394 has been selected as the major transport protocol, the 1394 CMM is mandatory in the HAVi architecture. The 1394 CMM provides two services to the HAVi system: a communication mechanism to control information exchange and an abstract representation of the network.

The IEEE 1394 is a dynamically configurable network where devices are identified via physical IDs. These IDs may dramatically change after the insertion or removal of a device in the network chain. The CMM combines the 24 bits of the IEEE 1394 vendor ID and the 40 bits of the chip ID to compose a permanent 64-bit global unique ID (GUID), which is independent of the IEEE physical ID. This GUID

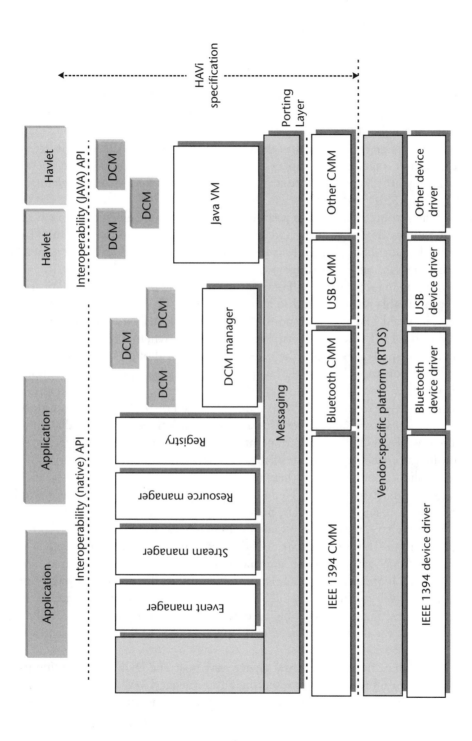

**Figure 10.4**   HAVi software architecture.

may be used for device identification, even if the device's IEEE 1394 physical address has changed.

The HAVi messaging system is independent of the Network and Transport layers and provides HAVi software elements with communication facilities. The messaging system of a device allocates software elements IDs (SEIDs), which are used to register software elements in the HAVi device register and for peer communication over the home network.

The registry is the system repository. It provides an API for software-element registration, management, querying, and retrieval throughout the home network. Any software element, before it contacts or is contacted by a peer element, has to register its SEID and attributes in the register.

The resource manager is responsible for the management of system resources and handles the reservation and release of hardware resources on the HAVI network. It also allows applications to register actions to be executed at a specific time (scheduled actions) and triggers them at the appropriate time. Finally, all of the resource managers on a HAVI network cooperate to ensure the best allocation overall of available resources and to avoid deadlocks.

The event manager is responsible for the event delivery service locally or globally in the HAVI network. An event may occur when a HAVi software element changes state. For example, the insertion or removal of a HAVi device may trigger an event. Software elements that want to be notified if a specific event happens register with their local event manager. When the event takes place, a software element posts the event to the local event manager, which distributes the event globally.

The stream manager supervises the continuous flow of multimedia data streams between HAVi devices. In cooperation with the resource manager, it allocates resources for the streams, establishes and maintains connections internally (within the appliance) or externally (between the appliances), and restores stream connections after a network break. To do this, the stream manager maintains a graph of all connections using their GUIDs, provided by the 1394 CMM.

The registry, messaging system, event manager, and resource manager are present on all full and intermediate A/V devices, while the stream manager is available on all full A/V devices and on intermediate devices only if it is required.

For each appliance, HAVi also provides an interface to the physical appliance, called a *device control module* (DCM). The DCM represents a device in the HAVi network and makes it available to external HAVi components. According to the category of the device, the DCM may operate in the following different ways:

- A full or intermediate A/V device may host one DCM representing itself and one or more DCMs representing base or legacy A/V devices. For example, an STB may be a full or intermediate A/V device hosting a DCM.
- A base A/V device may provide a DCM in JAVA bytecode. When it is attached to the HAVi network, the DCM is uploaded to a full A/V device and makes the

base device available to the HAVi components. If no full A/V device is available, but an intermediate device is able to handle it, the latter provides the DCM code to make the device available. For example, a video camera may be a base A/V device using the STB as a DCM engine.

- A legacy A/V device might not provide a DCM. When it is attached to the HAVi network, a full or intermediate A/V device that is able to handle it, provides the DCM code to make the device available. For example, a legacy VCR may be a legacy A/V device operating via the STB.

It is important to note that apart from the HAVi-specific layers and applications, a Java™ Virtual Machine (VM) is also supported on full A/V devices. The Java VM executes Java applications called *havlets*, which can be extracted from the DCM on request. Like applets, havlets are downloaded from a remote entity and then installed and run locally. While applets are downloaded from HTTP servers, havlets are downloaded from DCMs or application modules. In addition to providing an API for the controller to download the havlet, DCMs and application modules also provide an API that returns the havlet profile. The havlet profile is a description of the memory requirements of the havlet. By first examining the havlet's profile, a controller can estimate whether it has sufficient memory to run the havlet and the potential for havlet execution failure.

HAVi will make it easier for companies to build and market new application programs by using HAVi's APIs or programming in Java. Bridges or RGs will also be available for home control systems, security systems, and communication systems. As Java is becoming the major language for developing interactive networked applications on the Internet and for digital broadcasting, its support by HAVi will greatly benefit the user in getting access to interesting applications that will exploit the benefits of his HAVi network at home.

## 10.4  Universal Plug 'n' Play

Universal plug 'n' play (UPnP) [9, 10] is a higher-layer protocol stack that aims to extend the simplicity and autoconfiguration features of device PnP to the entire network, enabling discovery and control of networked devices and services [10, 11]. Built over the standard IP, HTTP, and extensible markup language (XML), UPnP enables a device to join a network dynamically, obtain an IP address, convey its capabilities, and learn about the presence and capabilities of other devices. Devices can automatically communicate with each other directly without additional configuration.

Home API was another standardization effort that supported multimedia home networks and allowed application programmers to access home devices. Home API allowed an application to query the status of home devices, but it did not support data streaming, like sending a video stream from a PC to a digital VCR. The founding members of Home API working group were Compaq, Honeywell, Intel, Microsoft, Mitsubishi, and Philips. Towards the end of 1999, Home API decided to merge its efforts with the UPnP Forum to ensure a unified specification for the development of home-control software and other products.

### 10.4.1  UPnP Network Architecture

UPnP is an open network architecture based on the principles, protocols, and applications of mature computer LANs. Leveraging the universal IP protocol, UPnP can be used over most physical media including radio frequency (RF, wireless), phone line, power line, coaxial, IrDA, Ethernet, and IEEE 1394. In other words, any medium that can be used to network devices together can enable UPnP.

As Figure 10.5 shows, UPnP is a pure IP network where UPnP devices are connected. However, other technologies (e.g., HAVi, CEBus, LonWorks, or X10) could be accessed via a UPnP bridge, proxy, or RG.

UPnP classifies the logical nodes into four categories:

1. *Control Point:* Intelligent, active, UPnP devices that host a set of software modules and are able to communicate and supervise a number of controlled devices. For example, a PC, laptop, PDA, RG, or advanced STB may play the role of a control point.
2. *Controlled Device:* Less intelligent, passive, UPnP devices able to respond to a control point and perform an action. For example, a DVD, VCR, or automated light controller may be a controlled device.

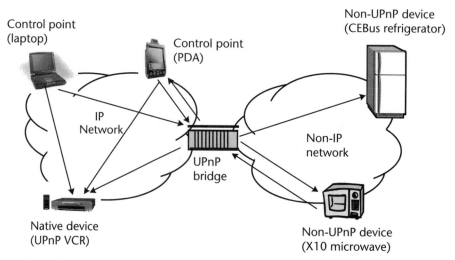

**Figure 10.5**  UPnP network architecture.

3. *UPnP Bridge:* Intelligent, multiprotocol, multitechnology UPnP devices that allow UPnP devices to communicate with legacy devices.

4. *Legacy Devices:* Devices that are not UPnP compliant or cannot participate in UPnP either because they do not have sufficient hardware resources or because the underlying media is unsuitable to run the TCP and HTTP protocols.

### 10.4.2   UPnP Software Architecture

To ensure interoperability between vendor implementations and to gain maximum acceptance in existing networked environments, UPnP leverages many existing, mature, standard protocols used on the Internet and on LANs. The basic protocols used to implement UPnP are summarized in Figure 10.6 [11].

- *IP:* IP has been for years the de facto standard for computer communications. UPnP leverages the IP protocol's ability to span different physical media and enable the direct use of mature protocols like TCP, User Datagram Protocol (UDP), Internet Group Management Protocol (IGMP), and Address Resolution Protocol (ARP), as well as core services such as Dynamic Host Configuration Protocol (DHCP) and Domain Name Service (DNS).
- *TCP/UDP:* TCP and UDP are the most frequently used protocols over IP. TCP is a connection-oriented protocol and requires that a connection be established between the transmitter and the receiver endpoints before

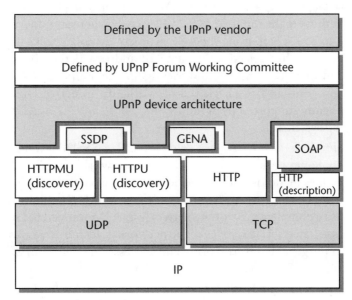

**Figure 10.6**   UPnP protocol stack.

communication is initiated. UDP is connectionless and the transmitter may send packets without the establishment of a connection.

- *ARP:* ARP is a member of the IP family of protocols. It is used to convert the IP addresses into physical MAC addresses. In UPnP, ARP is often used to determine if a particular IP address is or is not in use.

- *DNS:* For reasons of simplicity and categorization, all computers on the Internet are assigned a domain name. The DNS is an Internet/intranet service that translates computer domain names into IP addresses. DNS uses a distributed hierarchical database structure and delegates requests to local DNS servers to perform the name-to-IP-address mapping.

- *HTTP:* HTTP is the protocol used by the World Wide Web to transfer data between Web servers and browsers. HTTP is connection-oriented and uses the TCP protocol. When a query is received, the server opens a connection to the browser and sends the data. HTTP over TCP is also a core part of the UPnP Protocol, as UPnP has selected HTTP as the main method for data delivery.

- *HTTPU/HTTPMU:* HTTP over UDP (HTTPU) and HTTP Multicast over UDP (HTTPMU) are variants of HTTP, used when message delivery does not require the overhead associated with reliability.

- *AutoIP:* The default protocol for automatic assignment of IP addresses to computer systems is DHCP. The only drawback of DHCP is that it requires a server to assign the IP addresses. Initially, UPnP had selected DHCP. AutoIP is a proposal to the IETF that allows UPnP devices to obtain an IP address without a DHCP server. In the AutoIP proposal [12], the device selects an IP address in the range of 169.254.*xxx*.*yyy* where *xxx* and *yyy* are in the range of 1 to 254. Then, using the ARP protocol, AutoIP checks if another device in the network has selected the same IP address. If the address is already in use, AutoIP selects another and tries again. After an IP address is obtained, the system continues to check the network for an active DHCP server. AutoIP is already supported by Windows 98, 2000, and XP and by Mac OS 8.5 operating systems.

- *Simple Service Discovery Protocol (SSDP):* SSDP is a mechanism that enables home-network clients to discover network resources and services. SSDP defines methods both for devices to announce their availability and for control points to locate the resources on the network. As a result, every control point has a complete view of the network state. Moreover, SSDP enables devices and associated services to leave the network gracefully and includes cache timeouts to purge stale information. The successful result of an SSDP search is a ULR address. SSDP has been submitted as an Internet draft to IETF.

- *Generic Event Notification Architecture (GENA):* GENA provides the ability to send and receive notifications to subscriber entities using HTTP over

TCP/IP and multicast UDP. A control point interested in receiving event notifications will subscribe to an event source, and GENA will create the presence announcements to be sent using SSDP and to signal changes in the service state for UPnP eventing.

- *Simple Object Access Protocol (SOAP):* SOAP defines the use of XML and HTTP to execute remote procedure calls (RPCs). By making use of the Internet's existing infrastructure, it can work effectively with firewalls and proxies.

- *XML:* XML is "the universal format for structured data on the Web." It provides a methodology for storing structured data in text files. XML uses tags and attributes, just like hypertext markup language (HTML), but their interpretation depends on the context of their use.

- *UPnP Specific Protocols:* Based on the device architecture specification, UPnP vendors and working committees define specific UPnP protocols for device types, such as VCRs, refrigerators, dishwashers, and other appliances.

### 10.4.3   UPnP Operation

As previous sections have stated, UPnP provides basic network connectivity and addressing via a group of open, standard, Internet-based protocols, like the TCP/IP protocol suite and HTTP. On top of these, UPnP defines six important activities to be handled by HTTP servers, namely addressing, discovery, description, control, eventing, and presentation. In the next paragraphs, we are going to describe UPnP operation via these operations.

The prerequisite for all TCP/IP communications is the acquisition or definition of an IP address. In most LANs, a DHCP server is responsible for proper IP address assignment. In UPnP, if no DHCP server is available, the device may use the AutoIP protocol to get a temporary address.

After the attached device is assigned an IP address, it uses the SSDP module to advertise its services to the control points of the network. If the device is a control point, it additionally searches the network for devices and services. In both cases, a discovery message is exchanged, which contains device-related information, including the type and ID of the device along with a uniform resource locator (URL) to the device's description. The UPnP device description is an XML document that includes vendor-specific information, manufacturer name, model name and number, serial number, URLs to vendor-specific Web sites, and so forth. The description also includes a list of any embedded devices or services, as well as URLs for control, eventing, and presentation. The descriptions of the services are also written in XML and include lists of the commands, actions, parameters and arguments for each action, and lists of variables that represent the run time state of each service (e.g., data type, range, event characteristics). The UPnP control points use the device description to interact with the device.

Moreover, a control point may use SOAP and manage a device by sending a suitable control message to the control URL for the service. Control messages are also expressed in XML. The device responds with a message containing the status of the service, the specific values of an action, or a fault code.

When device or service documents change, the control points of the network have to be informed. For this purpose, the service publishes updates to the network by *eventing*. An event is an XML message formatted using GENA. It may contain the names of one of more state variables, their current values, and a list of actions and responses. All control points that have subscribed to these events will receive this update information. When a control point subscribes for the first time, it sends a special initial event message, which contains the names and values for all variables.

Finally, a device may have a page for device *presentation*. In that case, a control point can retrieve the page from its URL and load it into a Web browser. Depending on the page's capabilities, the user may control the device or view the device status.

As UPnP is based on standard protocols, APIs should enable six facets of UPnP: addressing, discovery, description, control, eventing, and presentation.

## 10.5    Open Services Gateway Initiative

The Open Services Gateway Initiative (OSGI) is an organization largely composed of equipment original equipment manufacturers (OEMs) and service providers, whose mission is to create open specifications for an end-to-end solution that enables the delivery of multiple services over WANs to home networks [13, 14]. OSGI is developing a services gateway specification that will enable service providers to "provide just-in-time value added services," while allowing the gateway to become a "service distribution, integration and management point in a SOHO or residence" environment [13].

The OSGI group was founded in March 1999 by the following companies: Alcatel, Cable and Wireless, Electricité de France, Enron Communications, Ericsson, IBM, Lucent Technologies, Liberate Technologies, Motorola, Nortel Networks, Oracle Corporation, Philips Electronics, Sun Microsystems, Sybase, and Toshiba. Soon, 22 more companies had joined, and OSGI became a conglomeration of companies from the software, hardware, and service-provider businesses. Some of the major additional members include AMD, Compaq, Deutsche Telecom, Berkom GmbH, GTE, France Telecom, HP, National Semiconductor GmbH, Nokia Corporation, Siemens AG, Schneider Electric, Sharp, ST Microelectronics, Telia Research, and Whirlpool Corporation. OSGI is speeding the adoption of its technology and accelerating demand for products and services based on these specifications worldwide through the sponsorship of market- and user-education programs.

### 10.5.1    OSGI Network Architecture

The OSGI specifications deliver an open common architecture for service providers, software and device vendors, system developers, and equipment manufacturers to develop, deploy and manage multiple services easily in a coordinated fashion [15].

The OSGI end-to-end network architecture shown in Figure 10.7 incorporates three key logical separated entities: the service and network providers, the services gateway, and the in-home network. While the figure shows a single services gateway, the OSGI architecture supports multiple services gateways, multiple WAN points of access, and multiple local networks working together to provide different services from different providers.

The service provider enables the provision of value-added services to the residential customer via the services gateway. Value-added services include downloading of software, application life cycle management, gateway security, attached device access, resource management, and functions necessary for remote administration of the gateway. For example a value-added service may be a home-security system, which includes physical-security-system monitoring sensors and activating alarms upon command from a remote server. If a service provider is trusted, secure download facilities ensure that only trusted code is installed on the gateway. A service aggregator is a service provider that offers the complete set of services that can be loaded into the gateway and is responsible for ensuring that all services are mutually compatible, and devices or resource requirements do not conflict.

The OSGI service operator plays a role very similar to the service aggregator's in that this entity manages and maintains the services gateway and its services; thus,

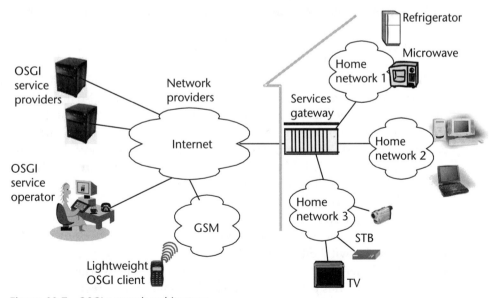

**Figure 10.7**    OSGI network architecture.

the terms are often used interchangeably. Examples of gateway operator functions include the following:

- Remote management of gateway resources and monitoring of device operating statuses;
- Remote control (download, start, stop, update, delete) of home-networking services;
- Definition and control of access rights between the gateway and the service providers and securing their communication channels;
- Control and management of the logical attachment of devices and local networks to the gateway, including authorizing the download of device and network drivers.

The network provider offers the necessary network infrastructure to enable the communications between the services gateway, the gateway operator, the service aggregator, and the service provider. This service may be provided and managed by a single wide area or metropolitan area carrier or via the Internet by an ISP. The network provider can perform other functions (i.e., telecom operator, wireless operator, energy-distribution company) and may or may not be the gateway operator. Although OSGI makes a clear distinction between the network provider and the gateway operator, combining these roles may turn out to be quite popular and profitable. For example, if one of the services provided by the gateway is Internet access and the network provider is an ISP, the separation between the network provider and the gateway operator may make it tricky to determine which is responsible for firewalls, access to files stored on home computers, routing, and other access issues.

The central component of the OSGI specification effort is the services gateway, which functions as the platform for many communications-based services. It will enable, consolidate, and manage voice, data, Internet, and multimedia communications to and from the home, office, or other location. It can also host a large variety of applications, such as energy management and control, safety and security services, health-care monitoring services, device control and maintenance, electronic-commerce services, and more. From the point of view of physical implementation, the services gateway will likely be integrated in whole or part in existing product categories, such as STBs, cable modems, routers, RGs, management systems, consumer electronics, or PCs.

The OSGI gateway will link many diverse home networks. Since the group's focus is on the Application Layer, OSGI claims to be medium agnostic. Thus, OSGI APIs will work with different underlying transport protocols, including SWAP, HomePNA, Ethernet, IEEE 802.11, and other various networking protocols. However, because OSGI is primarily Java based, it is naturally predisposed to interoperate best with Jini for internetworking connectivity within the Home LAN.

### 10.5.2   OSGI Software Architecture

To ensure a large target market for third-party service developers, as we ll as a large selection of compatible services for gateway operators, OSGI focuses on the services gateway protocol stack and specifies standard APIs for the platform execution environment. As Figure 10.8 shows, the OSGI software architecture consists of a group of components and modules, which operate over the platform operating system. Thus, the OSGI specification is hardware- and operating-system-independent. It specifies the following components and services:

- *Virtual Machine:* The OSGI execution environment operates over a Java™ Virtual Machine. The target is to allow implementations of services gateways to be based on lightweight versions of Java, with a very small footprint, special designed for consumer-electronic devices, such as Personal Java (pJava), Java 2 Micro Edition (J2ME), and other Java runtime environments.
- *Service Framework:* Over the Java™ Virtual Machine, a general purpose, secure, managed, service framework is defined, which is the core of the OSGI service-platform specification. The framework supports the deployment of downloadable applications called *bundles,* Java archive (JAR) file that do the following:
  - Contain the resources (e.g., Java classes) to support a service. Optionally, they may contain classes that help the service-maintenance framework (e.g., install, configure, activate, update, start, or stop a service).

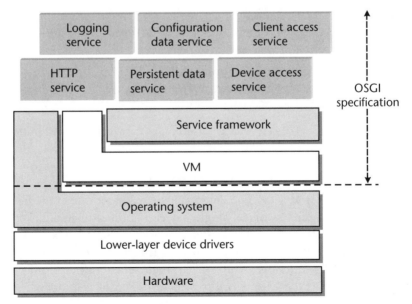

**Figure 10.8**   OSGI software architecture.

- Define static dependencies on other resources. If any dependencies are defined, the framework takes the required actions to make the relevant resource available.

- The advantage to bundles is that they can be uniformly downloaded and controlled. Moreover, they help maintain system consistency, as all dependencies can be handled uniformly. For example if a bundle is removed, all related services will automatically be unregistered.

- *HTTP Service.* HTTP has been one of the most widespread protocols worldwide; thus, OSGI specification includes an HTTP-based Web server, which runs on the services gateway device. Moreover, it defines an API, which provides facilities to configure the server, publish static content, and publish dynamic content generated from Java servlets. Servlets can be thought of as server-side applets, which extend the Web server's functionality.

- *Device Access Service:* The device access service is a mechanism that allows service providers to communicate with and control in-home appliances (e.g., alarm and lightening systems, consumer electronics, home PCs) via the services gateway without concern for low-level communication details. The device access service achieves this communication by dynamically loading on-demand specialized services for devices and network hardware connected to the services gateway. The features of the service include:

  - *Network and Device Independence:* It is open to any network architecture and device. New technologies and device types will simply require new device drivers.

  - *Automatic Detection:* It monitors the network and, wherever possible, locates and loads supporting device bundles with a minimum of user interaction.

  - *Legacy Support:* It supports older network technologies that do not support automatic device discovery and devices that do not contain OSGI-specific functionality.

- The device access service is based on two special types of bundles: the network bundle, which contains the necessary protocol stacks, drivers, and other resources to communicate with devices attached to the network, and the device bundle, which contains the code necessary to communicate with a specific type of device. The device access manager links together device and network bundles so that a single device bundle implementation can work with several different network bundles. For example, the USB network bundle contains the code to manage USB data transfer, detect new devices, and handle other USB functions, while only a single sensor device bundle is needed regardless of the type of network to which the sensor is attached.

- *Client Access Service:* The client access service is a mechanism that enables end users to view information in the gateway, modify configuration information, receive notifications, and interact with services in a uniform way. The client access service goes beyond the basic facilities provided by the HTTP service and provides a complete API for managing user interactions. Through data-format negotiation, it may provide access to clients hosted on devices that range from mobile phones to PC browsers.

- *Configuration Data Service:* The configuration data service provides a common service API for services configuration. The configuration may vary from authentication and authorization of information to services personalization. The configuration data service API enables configuration both locally by end users and remotely by the service operator or administrator.

- *Persistent Data Service:* The persistent data service provides a common mechanism for all services running on the services gateway to make information persist after the service's lifetime. It provides an API independent of the specific gateway operating system and file system, which enables services to store and retrieve persistent information, recover from errors and system failures, and synchronize the data with a server database. Implementations of the persistent data service may allow persistent information to be backed up on a server machine (perhaps one managed by the gateway operator) as an extra measure of reliability.

- *Logging Service.* The logging service is a specialized persistent data service that provides a logging mechanism. It defines an API that allows Java-based OSGI services to access a log file and read or write standard formatted information. During services gateway functional or operational troubleshooting, developers and operators may use this log file. The log service has the following general capabilities and features:

  - *Monitor Feature:* The logging service monitors framework events, creates log entries representing these events, and sends notifications for the created entries.

  - *Recording Feature:* This feature records in each log entry the system time, severity level, text message, and identity of the Java program that created the log entry.

  - *History Feature:* This feature can recover old log entries.

For more details on OSGI please refer to [13].

The OSGI approach is a step in the right direction, with a protocol stack that is quite broad and open, based on Sun's Java™ Virtual Machine technology. Unfortunately for service providers and public consumers, anytime either a Sun or Microsoft technology is fundamental to a standard's core, there will be strong opposition to the group and a lack of complete industry support. The opposing technology standard to OSGI would possibly be the Microsoft-led UPnP, although the aims of the two groups are completely different. UPnP aims to provide APIs above the Protocol Layer that facilitate the provision of services to devices using different protocols, while OSGI facilitates access to devices from an external network.

## 10.6  Versatile Home Network

Since 1995, the VESA home-network committee has been developing Versatile Home Network, or VHN, an interoperability standard that defines a digital in-home intranet that connects previously incompatible component or cluster networks. A component or cluster network is a homogeneous network of devices that reside together on a single physical network (e.g., a lighting-control network, A/V entertainment network, or an Ethernet data network for interconnecting PCs).

### 10.6.1  VHN Network Architecture

The VHN standard [16] specifies three types of networks and therefore three types of network interfaces (Figure 10.9):

1. Access-to-component interface;
2. Backbone-to-component interface;
3. Access-to-backbone interface.

External access networks include the PSTN/ISDN network and its xDSL variations, the cable and broadcast TV networks, satellite links, and air interfaces. All of these access networks are connected in a star topology with a VHN hub at the center of the star. The hub might be contained within an RG, STB, home server, or even a digital TV.

The VHN standard uses a long-distance version of IEEE 1394 (IEEE 1394b) as the digital backbone and IPs for interworking. Connectivity between the in-home component networks for both asynchronous and isochronous traffic is accomplished using IP over the VHN/IEEE 1394b backbone along with cluster-specific translators for asynchronous and isochronous communications between clusters. VHN also provides user-to-device control and device-to-device control across clusters and protocols.

### 10.6.2  VHN Software Architecture

Just like UPnP, VHN has selected and integrated a number of common, mature, and widely used mechanisms and protocols for network configuration and management. VHN uses the TCP/IP and UDP/IP protocol stacks as the base for device

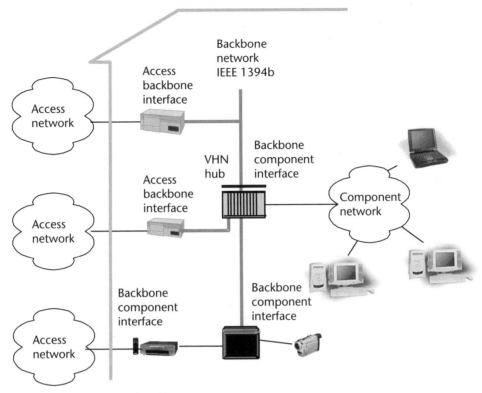

**Figure 10.9**   VHN network architecture.

communication, DHCP to allocate addresses, DNS for device namespace name-address mapping, and network address translator (NAT) for private global address mapping. Streaming data is supported using the Object Management Group (OMG) Media Streaming Framework, enhanced with A/V commands and HAVi to ensure that IEEE 1394–enabled A/V devices can be plugged directly onto a VHN backbone without conflicts.

User-to-device control is accomplished through a typical graphical user interface (GUI) proxy with direct support for HTML, while device-to-device control is based on a device-discovery query and a shared language, using hierarchical XML, RPC, and a home-network broker and interface repository (HNB and IR). The HNB and IR is a new concept introduced by VHN. It maintains the state of the network and device configuration information, allowing simple network maintenance and recovery of the network.

For network management, VHN uses the standard common information model (CIM) defined by the Distributed Management Task Force (DMTF). CIM applies an object-oriented approach to the management and control of devices, systems, and networks and uses common elements such as object classes, properties, methods, and associations. By incorporating the CIM standard, service providers will be able to design management and maintenance applications, which enable remote, cost-effective management and maintenance of a VHN-based home network.

VHN has many similarities to UPnP. In fact, VHN incorporates UPnP as a subset standard, thereby providing compatibility and interoperability between VHN- and UPnP-enabled devices. However, VHN goes beyond UPnP by specifying a Physical and Transport Layer using IEEE 1394, network management using the CIM, security based on Secure Sockets Layer (SSL) and IPSec, QoS for isochronous message transport, as well as a common interface to RGs that connect devices in the home to outside access networks.

## 10.7  European Home System

The European Home Systems Association (EHSA) is one of the major European propositions in home-systems standardization. EHSA is an open organization supported by major European electronic and electric companies, aiming to support and promote European industry in the field of home systems [17, 18].

To let electronic and electric devices from different manufacturers communicate with each other, EHSA developed the European Home Systems (EHS) specification based on the OSI reference model. The EHS specification defines the way electronic and electric devices in and around the home can interact and communicate with each other. Since 1987, major European industries have developed and validated EHS specification under the European programmers EUREKA and ESPRIT. EHSA's Standards Control Committee further developed EHS, involving also major European electrical utilities and semiconductor companies.

### 10.7.1  EHS Network Characteristics

The EHS specification defines a complete home-network system, which supports all domestic functions in a modular, extendible, and automatically configurable way. According to EHS, the home network constitutes a fully integrated control network built from one or more network sections, each using a single medium, linked together by routers.

Over this heterogeneous network, the EHS specification aims to provide a common LLC sublayer, suitable for all commonly available media, which enables the uniform transfer of control data, power, and information. EHS specification release 1.2 (Table 10.1) covers six media (twisted-pair type 1, twisted-pair type 2, coaxial, powerline, RF, and IR) to transport control data sharing the same LLC sublayer. Conformance and interoperability testing ensure that products from different manufacturers are interoperable.

**Table 10.1**   EHS Specification Network Characteristics

| Medium | Twisted-Pair Type 1 (TP1) | Twisted-Pair Type 2 (TP2) | Coaxial Cable | Powerline | Radio | IR |
|---|---|---|---|---|---|---|
| Target application | General purpose control | Telephony, ISDN, data, control | A\V, TV, data, control | Control | Telephone, control | Remote control |
| Bit rate | 9.6 Kbps | 64 Kbps | 9.6 Kbps | 2.4 Kbps | 1.2 Kbps | 1.1 Kbps |
| Access | CSMA/CA | CSMA/CD | CSMA/CA | CSMA/ACK | CT2 | — |
| Power feed | 35V | 35V | 15V | 230V ac | — | — |
| Channels | — | 14 | Many | — | 40 | — |
| Bit rate | — | 64 Kbps | Analog | — | 32 Kbps | — |
| Coding | — | TDM | FDM | — | FDM | — |
| Topology | Free | Bus | Bus | Free | Free | Free |
| Units/network | 128 | 40 | 128 | 256 | 256 | 256 |
| Range | 500m | 300m | 150/50m | House | 50–200m | Room |

## 10.8  KONNEX

KONNEX Association is the name given to the merger of the Batibus Club International (BCI), the European Installation Bus Association (EIBA), and EHSA into a common organization [19]. The technology of each of these associations was particularly suited to certain application areas, but none covered the full home-network range. The new association objective is to provide the technical basis for the convergence into a single common system supported by relevant industrial companies.

Several attempts worldwide to develop such a home and building electronic system have already taken place. However, the different system specifications have confused planning engineers, contractors, and installers as well as resellers, end users, building owners, and investors. This situation is hindering market acceptance and growth. Consequently, each system lacks the necessary volume of success. The newly established KNX technology aims to provide for the first time a common field bus platform suitable for all applications in the residential and building market.

## References

[1]   Ungar, S., "System and Architectural Requirements for a Broadband Residential Gateway. Request for Information," Bellcore, Bell Communications Research, July 24, 1997, pp. 33–37.

[2]   See http://sc25wg1.metrolink.com/ndocsold.htm.

[3]   ISO/IEC N617, Working Draft Home Electronic System, HES: Introduction, Draft Technical Report, ISO/IEC JTC1/SC25 WG1 N617, May 8, 1996.

[4]   ISO/IEC N710, HomeGate: Current Issues on Broadband Interworking and the Residential Gateway, Draft Technical Report, ISO/IEC JTC1/SC25 WG1 N710, June 16, 1997.

[5]  Zahariadis, T., Pramataris, K., Zervos, N., "A Comparison of Competing Broadband In-Home Technologies," *IEE Electronics and Communications Engineering Journal (ECEJ)*, August 2002, pp. 133–142.

[6]  See HAVI at http://www.havi.org.

[7]  Marsall, P., "Home Networking: A TV perspective," *IEE Electronics and Communications Engineering Journal (ECEJ)*, Oct. 2001, pp. 209–212.

[8]  HAVi, "HAVi, the A/V Digital Network Revolution," at http://www.havi.org/havi.pdf.

[9]  Miller, B., et al., "Home Networking with Universal Plug 'n' Play," *IEEE Communications Magazine,* Dec. 2001, pp. 104–109.

[10]  See UPnP at http://www.upnp.org.

[11]  Microsoft, "Understanding Universal Plug 'n' Play," white paper, 2000, at http://www.upnp.org.

[12]  Troll, R., "Automatically Choosing an IP Address in an Ad Hoc IPv4 Network," IETF Internet draft, March 2000, at http://www.ietf.org/proceedings/00jul/I-D/dhc-ipv4-autoconfig-05.txt.

[13]  See OSGI Home Page at http://www.osgi.org.

[14]  Marples, D., and Kriens., P., "The Open Services Gateway Initiative: An Introductory Overview," *IEEE Communications Magazine,* Vol. 39, No. 12, Dec. 2001, pp. 110–114.

[15]  Valtchev, D., and Frankov, I., "Service Gateway Architecture for a Smart Home," *IEEE Communications Magazine,* Vol. 40, No. 4, Apr. 2002, pp. 126–132.

[16]  EIA/CEA, "The VHN Standard," EIA/CEA 851 1.0, at http://www.global.his.com.

[17]  See the EHSA home page at http://www.domotics.com/homesys/Ehsa.htm.

[18]  See European Home Systems at http://www.smarthomeforum.com/ehs.shtml.

[19]  See Konnex Association at http://www.konnex.org.

# CHAPTER 11

# Residential Gateways

## 11.1   Introduction

It is expected that residential networks will rapidly increase network usage using a large number of new devices, ultimately improving the quality of life. The broadband digital home will provide residential users with integrated, value-added, multimedia services, such as VoD, music and news on demand, home shopping, distance learning, along with home control, security, and automation.

For years, the major obstacles to the deployment of the digital home were the lack of broadband access and in-home networks. Today, innovations in broadband access technology and huge investments in access infrastructures have brought the information superhighway to a critical mass of houses worldwide. Although, mature and emerging wireless and wired in-home-networking technologies clearly show that the technical barriers to residential broadband networking have been overcome, deployment of networked multimedia services to large audiences is still very limited. The major reason for this is the isolation of residential in-home networks. For the vast majority of consumers, electronic products are digital islands, and their functions cannot be coordinated across multiple devices or shared among multiple users.

The requirement for interoperability between isolated home network and interworking between different home appliances (e.g., PCs, printers, DVD players, TVs) over a home-network structure, along with the demand for high-speed Internet access, generated the need for a new device, the residential gateway, or RG. The RG is a single point of access and indoor network convergence, which will carry out the switching functions for telecommunications, computing, and entertainment service delivery, while providing overall control and management over a variety of electrical and electronic appliances.

This chapter will consider the RG's role in the end-to-end network architecture and its system requirements. Moreover, it will review some RG hardware and software architectures and discuss the RG business and ownership model.

## 11.2   The RG's Role

Since 1999, the term *RG* has been known only to a small group of industry pioneers who dreamed of creating a new product category that would enable the delivery of services into the home and onto the home network. Since then, the term has been used to describe any device that interconnects the access network with the in-home network, ranging from a simple modem to an integrated access device. Therefore, there is no clear definition of the term. In this book, we define the RG as a compact, scalable access platform that aggregates multiple services (voice, data, and Internet). It is the demarcation and interconnection device between the access and in-home networks, providing network termination, device interoperability, and service delivery.

Figure 11.1 shows the RG's position in future homes. The RG is located between the access network and the in-home network as a single convergence point and provides three main functions:

1. *Access Networks Connectivity:* The RG should be able to interface with the vast majority of broadband/narrowband wireless/wireline access networks. PSTN, ISDN, xDSL (ADSL, SDSL, VDSL), CATV, FTTH, Optical Ethernet, LMDS, MMDS, and satellite link are some potential RG access interfaces.
2. *In-Home-Network Concentration:* The RG should be able to concentrate all isolated in-home networks. Powerline, phone-line, Ethernet, USB, Firewire (IEEE 1394), and wireless protocols (IEEE 802.11b, HomeRF, Bluetooth, HIPERLAN2) are among the major in-home-network candidates.
3. *Networks Interworking:* The RG should be able to carry out the switching, routing, and interworking functions between any type of in-home or access network.

The RG should allow consumers to share Internet access between multiple in-home PCs, participate in multiplayer games or multipoint teleconferences, distribute video and music from any source (e.g., Internet, CATV, satellite, VCR, DVD, PC) on any appliance (e.g., TV, PC, game console) [1]. Moreover, the RG should enable interworking of different telephony systems and services—wired, wireless, analog, or IP based—and support telemetry and control applications, including lighting control, security and alarm, and in-home communication between appliances.

From the operator's and service provider's point of view, RG is a device able to multiply revenue in the end-to-end network architecture. By moving the service demarcation point into the customer's network, the RG enables the provisioning of integrated, value-added solutions, while minimizing the hassle of network configuration and management, as the RG can be managed remotely [2]. The RG may be a key element in the deployment of broadband services to residential users, as it may do the following:

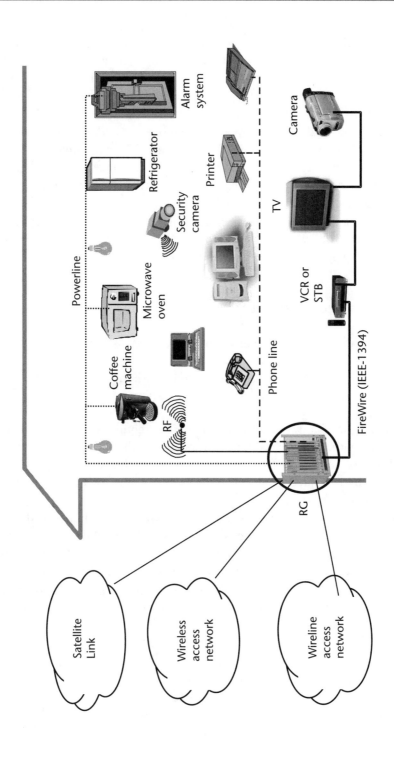

**Figure 11.1** RG positioning and role.

- *Lower Entry and Deployment Costs:* Operators can provide services with lower entry and deployment costs without the traditional hidden home-network costs associated with installation of wires and cables, network management, training, and the like. Moreover, scaleable and demand-based build-outs of RG systems, incorporating open industry standards, will ensure that services and coverage areas can be easily expanded as customer demand warrants.

- *Accelerate and Ease Systems and Services Deployment:* Systems can be deployed rapidly with minimal disruption to the community and the environment. Operators can offer two-way integrated voice, data, and video services over one network architecture without making any substantial changes to add new services.

- *Speed Up Realization of Revenues:* With the rapid deployment achievable via reusability of existing in-home wires and wireless technologies, service providers can expect earlier returns on their investments.

## 11.3   RG System Requirements

We have already emphasized that various types of RG systems are expected and that their functionality will dramatically differ. Of course, the requirements between the manifold RG models will vary significantly [3]; however, all RG systems are expected to have a number of basic characteristics:

- *Low Cost:* The first and most important requirement is the low cost of the final device. Some telecom operators require that RG prices be comparable to those for simple PSTN modems. Of course, the RG's complexity cannot be compared to such modems. However, customized high-volume production may really cut the price.

- *Multiple Flexible In-Home and Access Interfaces:* The RG will be the main point of convergence between the access and the home networks. Thus, it has to integrate multiple access and in-home-network interfaces and be flexible enough to accommodate additional emerging or future interfaces easily.

- *Voice/VoIP:* Voice is and will be a major requirement for residential users. Of course, solutions like VoIP are becoming more and more attractive, especially due to cost issues; however, the RG or the service provider should always provide interoperability with plain phone terminals.

- *Broadband Connectivity/Internet Access:* The majority of RGs are expected to have at least one high-speed connection to enable delivery of broadband services to the home. Delivery of integrated voice, data, and video will be required in the future. Moreover, the wide acceptance of the Internet has turned Internet access into a mandatory service.

- *Service Platform:* RG functionality is restricted to network interconnection. The RG is expected to be more than a multi-interface router. It is expected to be an intelligent and flexible service platform. It will be the heart and brain of the future digital home. It will provide overall control and management over a variety of electrical, electronic, telecom computing and home automation appliances.

- *Security/Firewalling:* The RG will interface with at least one public network. Moreover, different security requirements should be applied when the users surf the Internet or access the corporate Intranet. Thus, multiple layers of security should be provided to protect not only the RG, but the whole in-home network from malicious access attempts.

- *Robustness:* As the RG may be the single point of connectivity and communication, robustness will be quite important. Functions, such as autonomous operation, backup battery, and lifeline voice service will sooner or later be required. This is also the reason why solutions that combine a PC with a digital modem, although they may provide the basic functionality, are not considered RGs.

- *User Friendliness/Plug 'n' Play/Remote Management:* The RG targets the residential-user market. Thus, among the differentiation issues will be the level of user friendliness, which will ultimately provide fully PnP and autoconfiguration functionality. If management, troubleshooting, setup, or software upgrading is required, the RG should provide the operator or the service provider remote management capability.

## 11.4  Potential Hardware Architecture

Two approaches are envisaged to the RG hardware architecture: one modular and the other compact [4]. The modular architecture, shown in Figure 11.2, targets the SOHO and the small-business market. It consists of a switching backplane, which interconnects various peripheral feature cards (FCs). Each FC realizes a NIC and is responsible for Physical Layer transmission and control, analog-to-digital signal conversions and vice versa, protocol translations, and data communication with the backplane in a common data format. The backplane provides adequate switching capacity to support most existing or forthcoming indoor and access interfaces, while a controller FC provides the overall supervision and control. According to RG design and available backplane slots, the controller may be attached or integrated into the backplane. It controls the switching unit, coordinates the signaling communication between the FCs, supports advanced call handling and routing operations (e.g., load balancing between equivalent interfaces, overall system-resources management, rule-based session control), and provides for a single point of RG

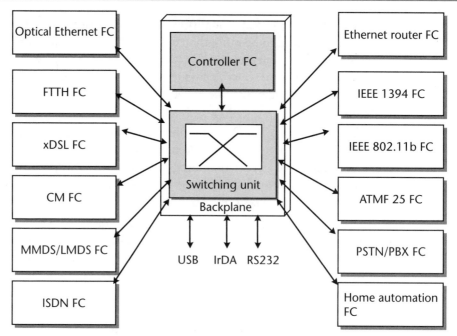

**Figure 11.2**  A modular RG hardware architecture.

management. Finally, a number of narrowband interfaces (e.g., RS232, USB, Bluetooth, IrDA) and input/output interfaces (e.g., LCD display, console, IR/RF remote control) may be directly attached either to the backplane or to the controller FC.

The design philosophy behind the modular RG architecture is to create an expandable RG that can accommodate any viable interface or protocol by transferring the complexity of the specific protocol and the peculiarity of the physical medium to the peripheral interface FCs. Data is transferred transparently via the switching fabric, while control and operation administration and management (OA&M) messages are supervised by the control FCs; if required, they are forwarded to the appropriate FC for processing. As a result, whether the FC is simple or quite complex, the RG is able to support it. Moreover any new, emerging, or evolutionary interface can be introduced on-demand, according to user needs or service provider offerings.

Figure 11.3 shows a modular architecture RG prototype and some FC interfaces (by Ellemedia Technologies, Ltd.). The prototype integrates a high-speed switching backplane with up to 2.5-Gbps aggregate switching capacity, a control processor, and some basic interfaces (i.e., RS232, USB, IrDA, PC card/IEEE 802.11b, console). Any combination of interface FCs (e.g., Switched Ethernet, CEBus, IEEE 1394, PSTN/PBX, ATMF-25, ATM OC3, ISDN) may be connected to the eight equivalent

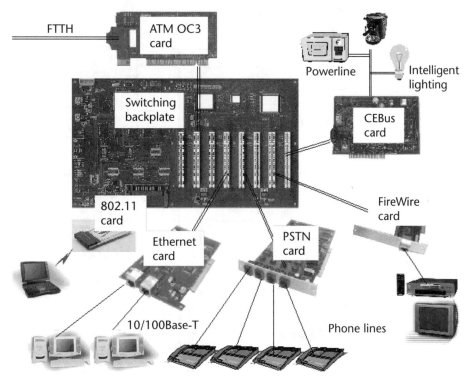

**Figure 11.3**  Modular RG prototype example. (*Source:* Ellemedia Technologies, Ltd.)

slots and can be used as an access or indoor interface. PnP and OA&M functions on the control processor enable overall RG operation [5].

Although the modular design is very flexible and expandable, it may lead to a quite expensive and complex multiprocessor system. As an alternative, a compact RG architecture is shown in Figure 11.4. The compact RG targets only the residential-user market. In the compact RG version, a set of modules and interfaces is considered mandatory or default. This set includes the RAM and flash memories, some standard interfaces such as USB, RS 232, and Ethernet, and at least one access interface. Vendor differentiation will mainly be based on the processor speed, system functionality, cost, selection of the additional modules (e.g., hard disk drive, DVD/MP3 player, monitor interface), and the access and in-home-network interfaces (e.g., IEEE 802.11b, powerline, DOCSIS, xDSL, Optical Ethernet). Of course, the compact RGs are less modular and scalable; however, they are also cheaper and this is a very important factor in the consumer-electronics arena.

In Figure 11.5, we present a compact RG prototype with multiple access and indoor interfaces (by Ellemedia Technologies Ltd.). Of course, due to performance and cost issues, only a subset of these interfaces may be selected.

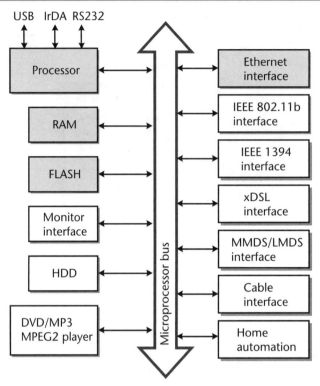

USB  IrDA  RS232

**Figure 11.4**  Compact RG hardware architecture.

Another important issue is RG operation in the event of power outage. Two approaches can be considered. The first is based on the assumption that most indoor appliances and access networks will also power down just after the outage; thus, it is unnecessary to have an active RG, although some exceptional operations (e.g., emergency telephone calls) may be covered by designing specialized FCs (e.g., PSTN lifeline). The second approach assumes there is a backup battery or UPS support; thus, power-autonomous devices (e.g., laptops) will continue to operate normally, at least without loosing indoor network connectivity. Both approaches are reasonable, and the selection of one may depend on the RG target market, services, and cost.

## 11.5  Potential Software Architecture

Besides the flexibility and modularity of the RG hardware architecture, the variety of access and indoor network interfaces, and the marketing, installation, and maintenance costs, the success of the broadband digital home strongly depends on the

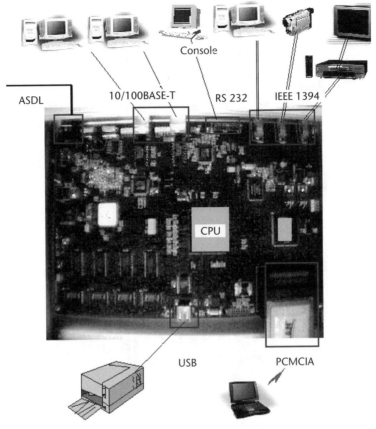

**Figure 11.5**   Compact RG prototype example. (*Source:* Ellemedia Technologies, Ltd.)

efficiency of the RG software architecture and the user-friendliness of the applications. Significant efforts and open standards for interoperability and interoperation already exist (e.g., OSGI, HAVi, HES, UPnP). These standards were described extensively in Chapter 10.

In this section, a software architecture that adapts the UPnP standard to the requirements of an embedded multi-interface RG platform is presented. The proposed RG software stack is shown in Figure 11.6. It takes into account the limited hardware capabilities of the RG device (memory size, processing capability, storage capacity) and the requirement for minimal interaction with the user. Taking a bottom-up approach, the lower layer just over the system hardware consists of the software that interfaces with the following hardware system resources:

1.  The on-board device drivers that control the on-board peripherals (e.g., RAM, flash, console, LCD display) and interfaces (e.g., RS-232, USB, IrDA);

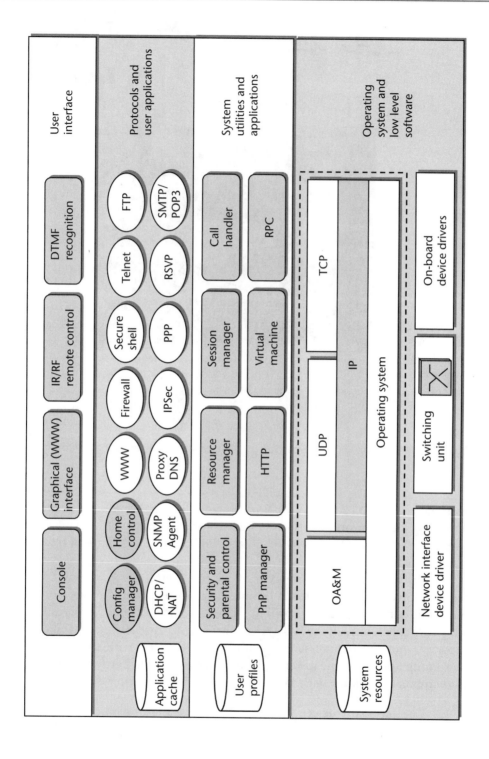

**Figure 11.6** Proposed RG software architecture.

2. The network device drivers that control the network interfaces;
3. Optionally, the switching module that initiates, monitors, and controls the switching unit.

The embedded operating system acts as a glue, creating the illusion of seamlessness for application programs. Some OS characteristics include small footprint and size, reduced memory requirements, portability, and reliability, while real-time, QoS, and security extensions may be expected from modern embedded operating systems.

Over the operating system, the IP is the dominant layer, accompanied by TCP and UDP, while some OA&M hooks are left for applications that require direct access to the system kernel. In this layer, the system resources database is also stored in the form of a management information base (MIB) or system registry. This database contains all system-related information, including available interfaces and slots, technical characteristics (e.g., physical medium, bandwidth, protocols, processor, memory, MAC addresses), and all session-related information.

The System Utilities and Applications Layer accommodates the RG's system-specific applications and operating-system-related protocols and utilities. According to RG functionality, modules may be included, adapted, or omitted. The operational and control distribution requirements indicate the level of abstraction, which may vary from simple RPCs or a VM to a complete distributed piece of middleware. The RG PnP module provides functionality similar to that of the UPnP SDP. It detects the system and network resources, while the resource manager coordinates and keeps the home-resources database consistent. Assisted by the PnP module, the resource manager may also provide for personal mobility by seamlessly detecting mobile and wireless terminal and handheld devices located in the wireless-home-network area. After identification, the home devices are authenticated by the security and parental control module, which is also responsible for all user authentication and authorization operations, user permissions and parental control checking, and supervision of call-and-service requests. The security module may also control data-stream decoding or encryption and decryption and provide cryptography functions, codes, and keys. User permissions and preferences are stored in the user profiles database, which is updated by the OA&M module and higher-layer applications.

The call handler and session manager are also important modules for session-oriented applications. The call handler sets up any-to-any connections between input and output devices, servers, adjunct processors, and the like. It may bridge, route, or switch command, control, or data flows between subsystems and media and define fallback procedures for applications with mismatched terminals or servers. Intelligent multimedia call handlers may also provide intercommunication or PBX functions for (IP/H.323) voice and video telephony, handle teleconference bridging, and handle call-by-call requests based on estimated or predefined "best" access options, prices, or user or service preferences. Finally, the session manager keeps track of session requests, monitors RG system resources, suspends and reactivates application sessions, and calculates accounting and billing information.

The Protocols and User Applications Layer contains the protocols and the applications that provide the RG's higher-layer functionality. In Figure 11.6, we provide a set of potential RG applications, which may be organized into two groups: the supporting protocols and utilities (e.g., DHCP, NAT, RSVP, PPP, POP3/IMAP) and the autonomous applications (e.g., WWW, Firewall, Telnet, Secure Shell, FTP). Also, some applications, like the configuration manager described in the next paragraph, and the home control, which provides for home automation and security, are specialized RG applications.

Finally, the RG may support a variety of user interfaces for operation, control, and management purposes. A nonexclusive list may include powerful user interfaces, such as a console or a Web-based GUI, or interfaces with reduced functionality like DTMF for RG control via a phone line, IR, or RF remote control, and the like.

## 11.6   Business Model

One of the most critical elements of the deployment of RG systems is the ownership model. Generally, we may identify the following ownership cases:

- *End-User Ownership:* In this case, the cost of RG system acquisition and installation is covered solely by the end user, who has the right to install, upgrade, or remove the equipment and is free to select the best access provider, network provider, and service provider.
- *Service/Network Provider Ownership:* In this case, the provider offers the RG system, and the user just pays for the services and the content. This approach fits quite well in cases where the provider owns the access network and RG intelligence and the cost is rather limited.
- *Subsidized Cost:* In this model, the user buys the RG and the required interfaces at a much lower price subsidized by the provider. The RG equipment belongs to the user, but he has to buy long service contracts from the respective provider. The basic monthly cost makes up for the provider's initial investment. The user, on the other hand, can buy the service without making the initial upfront investment.
- *Mixed Ownership:* Especially for highly modular RG systems, a mixed model of ownership is quite interesting. According to this method, the basic RG configuration is bought by the end user or acquired at a subsidized price, and the extra cost for each additional interface or service is covered by the end user or the provider, depending on the service or interface.

Among the ownership models, the promotional and subsidized-cost approaches seem most realistic:

- They do not burden either the end user or the provider unilaterally.
- They minimize the initial cost from the end user's point of view.
- They bind the customer to a specific network or service provider.

Of course, depending on the respective market and social situation, any service combination can be regarded as meaningful enough for the subsidizing of the associated interfaces.

## 11.7   Market Forecast

The home-network market is expected to be valued at $5.5 billion by 2005, with 68 million households having multiple computers and about 37 million households having some type of home-networking connection, according to Parks Associates. As shown in the forecast by In-Stat/MDR in Figure 11.7, RGs represent an emerging market worldwide. It is important to emphasize that not only the U.S. market, but also European and Asian markets, will be ready for broadband services in a couple of years. Moreover, RGs are expected to play a critical role in home networks, mainly because they offer great opportunities for the operators and service providers to provide end-to-end broadband services to residential users at home.

For many service and hardware providers, this translates into even greater profits. Many vendors have already started to deliver RGs with various interfaces and

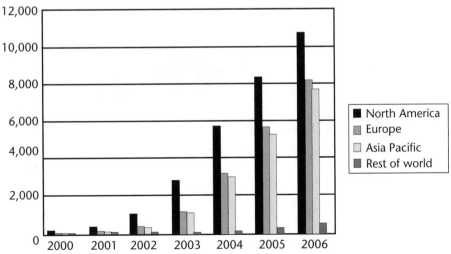

**Figure 11.7**   RG regional forecast ('000s). (*Source:* In-Stat/MDS, 2001 [6].)

functionalities. Here, we present a very limited representative subset and group it into two approaches. Figure 11.8 shows samples from manufacturers that have launched powerful and more expensive RG products. They are mainly small routers or extended STBs with multiple interfaces that target the small-business, SOHO, and residential-user markets.

Figure 11.9 shows samples from manufacturers that have launched low-power and low-cost RG products. They are specially designed for the residential-user market, and they have a small number of basic interfaces.

The market will finally show which direction is more appropriate. However, it seems that more powerful RG products with a large number of interfaces and high integration to lower the cost would be more successful [7, 8].

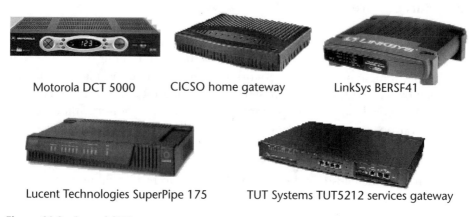

**Figure 11.8**   Powerful RG systems.

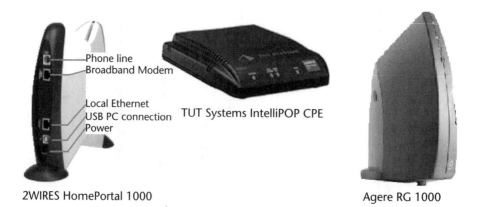

**Figure 11.9**   Low-end RG systems.

# References

[1]    ISO/IEC, "Architecture of the Residential Gateway," ISO/IEC JTC SC25 WG1 N912.

[2]    Teger, S., and Waks, D., "End-User Perspectives on Home Networking," *IEEE Communications Magazine,* Vol. 4, Apr. 2002, pp. 114–119.

[3]    Ungar, S., "System and Architectural Requirements for a Broadband Residential Gateway. Request for Information," Bellcore, Bell Communications Research, July 24, 1997, pp. 33–37.

[4]    V. Lawrence, et al., "Digital Gateways for Multimedia Home Networks," accepted for publication in *Telecommunication Systems Journal,* August 2003.

[5]    Zahariadis, T., Pramataris, K., and Zervos, N., "A Comparison of Competing Broadband In-Home Technologies," *IEE Electronics and Communications Engineering Journal (ECEJ),* Aug. 2002, pp. 133–142.

[6]    Wolf, M., "The Digital Domicile: The Exploding Market for Home Networking Technology and Services," Cahners In-Stat Group, Oct. 2001.

[7]    ISO/IEC N738, "HomeGate: A Residential Gateway Model for HES," ISO/IEC JTC1/SC25 WG1 N738, Feb. 16, 1998.

[8]    ISO/IEC N710, "HomeGate: Current Issues on Broadband Interworking and the Residential Gateway," Draft Technical Report, ISO/IEC JTC1/SC25 WG1 N710, June 16, 1997.

# CHAPTER 12

# The Present and Future of Home Networking

The vision for the future digital home includes a computer in each and every house. Twenty years ago, this prediction would not have been far from science fiction. Today, computers have become an integral part of our lives. The majority of jobs requires at least a basic knowledge of computers, while the number of telecommuters working from home rapidly increases. Most houses in Europe are equipped with at least one PC, and more than half of the houses in the United States have two. Moreover, the (r)evolution of embedded microprocessors has generated a new generation of intelligent consumer-electronic devices. Intense competition between manufacturers has maximized microprocessors' capabilities and minimized their size, while large volume production has minimized their cost. As a result, more and more home appliances are powered by a microprocessor: in the TV to control the contrast and volume, tune the receiver, store channels, handle teletext; in the VCR to store the automatic recording program and the user preferences; in the hi-fi to decode MP-3 songs; in the washing machine to provide fuzzy logic and on-site upgrade capabilities. Finally, from an academic network the Internet has turned into a de facto communication medium.

Interconnecting home PCs, consumer-electronic devices, and home appliances in a home network represents a unique opportunity to set the ground for a vast range of new home applications and value-added services. Moreover, access to the information superhighway from home expands the customer base beyond the saturated corporate environment to residential users. This new, potentially huge market and the increasing number of network-aware, intelligent appliances have driven vendors to develop innovative, broadband access- and indoor-network technologies, and telecom operators to proceed with enormous investments to bring the information superhighway into residential users' homes.

For years the inadequate access-network infrastructure and the huge cost of new installations have hindered the deployment of the digital networked house. The notion was that either the telephone or CATV companies would provide the last mile of service. Today, a number of emerging access techniques, ranging from

CATV, copper enhancements (DSL), and fixed WLL to satellite and FTTH have already been tested and evaluated, and their large deployment is already underway.

In the arena of indoor networking, multiple current and emerging wireless and wired technologies over multiple existing or future network architectures and physical media aim to provide multimedia home systems (inter)communication.

In this book, we have studied a range of deployed and emerging home-networking technologies with wiring, no-rewiring, or no-wiring requirements. Among the no-rewiring technologies, the powerline technologies use the most common in-home network, but they suffer from noise and interference due to the powerline network or wiring environment. Phone-line technologies have superior performance due to better their network architecture, but they are limited by the number and location of phone jacks in the house. Wherever available, the coaxial network with enhancements is also a very good solution for carrying higher-layer protocols.

Among the technologies with no-wires requirements, which are expected to dominate in the home-network domain, IEEE 802.11b is the most widespread and mature wireless solution, while emerging technologies (e.g., IEEE 802.15.3, HIPERLAN2) aim to provide enhanced features. For short-distance, low-cost, and limited-power consumption, technologies like Bluetooth are envisaged. From the technologies that need house rewiring, IEEE 1394 is the proposed standard, while 10/100BASE-T UTP-5 Ethernet will be used especially for home offices.

The book also presented a reference in-home-network architecture and introduced the RG modular and compact architectures. It is important to note that apart from the physical medium and the RG's functionality, interoperability between devices from different manufacturers has to be achieved at higher layers. HAVi, VHN, UpnP, and EHSA are among the candidate higher-layer protocol stacks expected to provide for in-home device interoperability.

The success and wide acceptance of the above solutions heavily depends on their performance, reliability, simplicity, and usability, as well as the cost of the home telecommunications systems and of network installation, operation, and management as a whole. Services will only be able to attract new customers if they can provide the appropriate functionality and flexibility, fulfill user requests, guarantee the agreed-on quality, supply sufficient content, and compare favorably with standalone systems.

The path towards broadband home networks incorporates a number of critical trends. By definition, it is difficult to make precise predictions and foresee which systems, solutions, and services will be considered successful after 10 years. An important factor in uncertainty is the limited deployment of interactive TV systems and rather unsuccessful trials with residential users. However, two things can be taken for granted:

1. Future home networks will be deployed in an environment where many wired and wireless network-aware consumer-electronic appliances will already be deployed. The requirements for interoperability between independent home-network segments, physical media, transport protocols, and applications should be considered mandatory.

2. Operators have made huge investments in access-network infrastructures. Future access networks will have to provide really break-through technologies and services to persuade investors of their necessity. Thus, home networks should be able to interoperate with existing and emerging access networks.

Home networking represents a quickly growing arena. After deployment of current and emerging technologies, new applications will appear, turning digital networked houses into smart adaptable homes. More intelligent and autonomous consumer-electronic devices are expected to communicate with each other, take their own initiatives, and handle specific every-day tasks without human intervention. Smart networked automation and home-security systems will provide advanced services and secure the house from all kinds of thugs, hooligans, and terrorists. Broadband in-home networks will interconnect devices, provide adequate bandwidth capabilities, and hide any heterogeneity in home segments. Intelligent RGs will coordinate the smart house, interface with legacy systems, and provide value-added services. Science fiction films are expected to be reality soon. As they say, the only limit is the human imagination.

# Glossary

**1394**  An IEEE high-speed serial bus standard for transferring audio-video content. It is also called *firewire*.

**5-UP**  A single standard developed by the ETSI and the IEEE, which aims to merge the 802.11a and HIPERLAN2 standards.

**802.11**  An IEEE high-speed wireless technology in the 2.4-GHz frequency range with a maximum 2-Mbps data-transfer rate.

**802.11a**  An IEEE specification for wireless networking in the 5-GHz frequency range with a maximum 54-Mbps data-transfer rate.

**802.11b**  An IEEE specification for wireless networking in the 2.4-GHz frequency range with a maximum 11-Mbps data-transfer rate.

**802.11d**  An IEEE task group that works towards 802.11b versions at other frequencies for countries where the 2.4-GHz band is not available.

**802.11e**  An IEEE task group that works towards the specification of a new 802.11 MAC Protocol to accommodate additional QoS provision and security requirements over legacy 802.11 Physical Layers.

**802.11f**  An IEEE task group that aims to improve a handover mechanism that will enable roaming between access points attached to different 802.11 networks.

**802.11g**  An IEEE task group that aims to define a new standard, fully compatible with 802.11b, but with 22-Mbps bandwidth.

**802.11h**  An IEEE task group that aims to enhance 802.11a to gain acceptance by the European regulators.

**802.11i**  An IEEE task group that aims to enhance 802.11 security.

**802.11j**  An IEEE task group that will propose the issue of 802.11a and HIPERLAN2 interworking.

**802.15.3**  An IEEE wireless specification that supports ad hoc networking, multimedia QoS guarantees, and data rates from 11 to 55 Mbps.

**ACF**   Association control function.

**ADSL Lite**   A lower-speed ADSL variation capable of providing rates of up to 1.5 Mbps downstream and 512 Kbps upstream. Also called or UADSL or G.lite.

**ADSL**   Asymmetric digital subscriber line is a DSL version with a range of up to 18,000 ft that transmits over a single copper twisted-pair cable at upstream rates of up to 640 Kbps and downstream rates of up to 12 Mbps.

**ANSI**   American National Standard Institute.

**Bluetooth**   A short-range wireless interconnection technology specification for hand-held devices like Palms, pagers, mobiles, PDAs, and the like.

**BPSK**   Binary phase shift keying.

**BRI**   Basic-rate interface is an ISDN service supporting up to 128-Kbps bandwidth or up to two voice, fax, or data parallel sessions.

**Cat 5**   Category 5 or Cat5 cable is twisted-pair copper cables rated for 10-Mbps and 100-Mbps date rates used for Fast Ethernet or 10/100 Ethernet.

**CEA**   The Consumer Electronics Association is a sector of the American National Standard Electronics Industries Association (EIA).

**CEBus**   Consumer-electronic bus PLC technology transports messages between household devices, using the home's 120V ac electrical wiring in a standardize way. The CEBus Protocol is an American National Standard (ANSI/EIA-600).

**CM**   Cable modem.

**CMTS**   Cable modem termination system.

**CPE**   Consumer premises equipment.

**CRC**   Cyclic redundancy check.

**CSMA/CA**   Carrier Sense Multiple Access with Collision Avoidance is a protocol for managing traffic on wireless networks (e.g., 802.11).

**CSMA/CD**   Carrier Sense Multiple Access with Collision Detection is a protocol for managing traffic on an Ethernet network.

**DBS**   Direct broadcast satellite uses GEO satellites for TV distribution.

**DCSA**   Dynamic channel selection and allocation.

**DECT**   Digital Enhanced Cordless Telecommunications is a digital radio access standard for cordless communications and de facto for voice communications.

**DHCP**   Dynamic Host Control Protocol dynamically assigns IP addresses from a predefined list to network devices.

**DNS** Domain Name System is a program that translates URLs to IP addresses.

**DOCSIS** Data over Cable Specifications is the most widely adopted standard for bringing multiple TCP/IP sessions and Internet access to subscribers over CATV networks.

**Downstream** Data flowing on a network traffic path from a service provider to an end user.

**DSL** Digital subscriber line (or digital subscriber loop) is a dedicated point-to-point public network-access technology for transporting broadband streams over plain copper telephone lines.

**DSLAM** DSL access multiplexer terminates the ADSL lines and splits voice from data traffic.

**EHS** European Home Systems is a specification based on the OSI reference model and developed by EHSA.

**EHSA** The European Home Systems Association is an open organization supported by major European electronic and electric companies, aiming to support and promote European industry in the field of home systems.

**Ethernet** International standard networking technology for wired implementations with a speed of 10 Mbps. Ethernet successor technologies provide up to 100 Mbps or even 1 Gbps.

**ETSI** European Telecommunications Standards Institute.

**Fast Ethernet** An international networking technology standard for wired implementations with a speed of 100 Mbps.

**FCC** Federal Communications Commission.

**FDM** Frequency division multiplexing.

**FDMA** Frequency division multiple access.

**FEC** Forward error correction.

**FEXT** Far-end crosstalk is an interference type, which occurs when two or more signals transmitted in the same direction on different UTP pairs have overlapping spectra.

**Firewall** A software system that prevents network access by unauthorized users.

**FireWire** The IEEE 1394 standard.

**FP** A DECT fixed station that hosts one or more base stations (Radio Fixed Point, or RFP).

**FTTB**    Fiber-to-the-building extends the fiber-optic infrastructure to corporate or apartment buildings.

**FTTC**    Fiber-to-the-cabinet or -curb extends the fiber-optic infrastructure to the street cabinets of a neighborhood.

**FTTH**    Fiber-to-the-home extends the fiber-optic infrastructure to the home.

**GEO**    A geostationary Earth orbit satellite appears at a fixed latitude and longitude at an altitude of around 36,000 km.

**HAVi**    Home audio video interoperability is a standard that aims to allow all manner of digital consumer electronics and home appliances to communicate with each other.

**HDSL**    High bit-rate digital subscriber line is a symmetric DSL service, providing 1.544 Mbps over two copper pairs and 2.048 Mbps over three pairs.

**HDTV**    High-definition television.

**HEC**    Header error control.

**HES**    Home Electronic System is an international standard for home automation; it is part of the ISO/IEC.

**HFC**    Hybrid fiber-coaxial is an access-network technology that combines optical fiber and coaxial CATV in different portions of a network to carry broadband content at home.

**HIPERLAN**    High-Performance Radio LAN is the European reply to the IEEE 802.11 standard.

**HIPERLAN2**    The European proposition for a broadband wireless LAN operating with data rates of up to 54 Mbps at the Physical Layer on the 5-GHz frequency band.

**Home RF**    A short-range wireless in-home interconnection technology.

**HomeCNA**    The Home Cable Network Alliance is an industrial alliance aiming to standardize the physical aspects of the home coax network.

**HomePlug**    A networking industry group of companies working toward the standardization of specifications for powerline networking products.

**HomePNA**    The Home Phoneline Networking Association is a consortium of companies working on technologies that will enable transmission of high-speed data over the in-home phone-line network.

**IDSL**    ISDN digital subscriber line is a symmetric DSL with ISDN speed. It provides full-duplex data-only service throughput at speeds of up to 144 Kbps.

**IEC** The International Electrotechnical Commission is part of the ISO.

**IEEE** The Institute of Electrical and Electronic Engineers is an international organization that sets electronic and electrical standards.

**IP** Internet Protocol is the most widely accepted communications protocol worldwide.

**IrDA** The Infrared Data Association is an international organization that develops technical standards for electronic data exchange via infrared technologies.

**ISDN** Integrated services digital network provides multiple voice and digital data services over regular phone lines.

**ISO** The International Organization for Standardization is perhaps the largest standardization organization worldwide.

**ISP** An Internet service provider is a company that provides Internet access to individuals and businesses.

**ITU** The International Telecommunications Union is a global organization whose mission is to adopt telecommunications regulations and standards.

**KONNEX** This association is the result of the merger between the Batibus Club International (BCI), the European Installation Bus Association (EIBA), and the European Home Systems Association (EHSA).

**L2CAP** Logical Link Control Adaptation Layer Protocol.

**LEO** A low Earth orbit satellite rotates at an altitude of around 500 to 2,000 km.

**LMCS** Local multipoint communications system. See LMDS.

**LMDS** The local multipoint distribution service, also referred as the local multipoint communications system (LMCS), is a broadband, fixed-wireless access system, which allows for two-way digital communications.

**MEO** A medium Earth Orbits satellite rotates at an altitude of around 10,000 km.

**MMDS** The multichannel multipoint distribution service, also referred as wireless cable, is a fixed-wireless technology that has been used primarily for analog TV broadcasting.

**MPEG** The Moving Picture Experts Group is the ISO/IEC group that has defined the MPEG-1, MPEG-2, MPEG-4, MPEG-7, and MPEG-21 standards. MPEG-1 and MPEG-2 are Emmy Award–winning standards that made interactive video on CD-ROM and digital television possible.

**MPEG-1** An ISO/IEC encoding standard for interactive video distributed on CD-ROM.

**MPEG-2**   The most famous ISO/IEC encoding standard for interactive digital TV distribution.

**MPEG-4**   An ISO/IEC video-encoding standard, which targets video distribution over the Internet and mobile networking applications.

**NAT**   Network address translation is a mechanism that allows for the dynamic reuse of an IP address for all PCs on a network.

**NEXT**   Near-end crosstalk is an interference type that occurs when there is a spectral overlap between signals traveling in opposite directions on different pairs in a UTP cable.

**NIC**   A network interface card is a type of PC adapter card that provides two-way communication between network devices.

**NT**   Network termination.

**OLT**   Optical line termination.

**ONT**   Optical network termination.

**ONU**   Optical networking unit.

**OSGI**   The Open Services Gateway Initiative is an organization comprising equipment OEMs and service providers, which creates open specifications for the delivery of multiple services to home networks.

**PLC**   The Powerline Carrier standard aims to deliver burst data rates of up to 20 Mbps over powerline cables.

**PnP**   Plug 'n' play is a computer-system feature that provides for automatic configuration of add-ons.

**PON**   Passive optical network.

**POTS**   The plain old telephone service is the standard analog telephone service.

**PP**   Portable point is the terminal in DECT architecture/terminology.

**PRI**   The primary-rate interface is a broadband ISDN configuration supporting up to 2 Mbps (E1) in Europe, Central and South America, and Asia, or 1.5 Mbps (DS1 or T1) in the United States.

**PSTN**   The public-switched telephone network is the public telephone network.

**QOS**   Quality of service.

**QPSK**   Quadrature phase shift keying.

**R-ADSL**   Rate-adaptive digital subscriber line is an adaptive ADSL variation, where the transmission rate may be adjusted dynamically.

**RAM**   Random access memory.

**RFI**   Radio frequency interference.

**RFP**   The radio fixed point is the base station in DECT architecture/terminology.

**RG**   A residential gateway is a network device that interconnects the access and the in-home networks.

**RJ-11**   A phone-line connector used to connect a phone or a modem to a phone line.

**RJ-45**   An eight-pin connector for Ethernet cables.

**RLC**   Radio link control.

**ROM**   Read-only memory.

**RRC**   Radio resource control.

**SAR**   Segmentation and reassembly.

**SDSL**   Single-line digital subscriber line is a symmetrical DSL service at ISDN PRI rates, but it uses a single copper-pair wire over distances of up to 22,000 ft (6.7 km).

**SOHO**   Small office home office.

**SWAP**   The Shared Wireless Access Protocol is a standard used in HomeRF.

**TA**   Terminal adapter.

**TCP/IP**   The Transmission Control Protocol/Internet Protocol is the default standard Internet communication protocol.

**TDM**   Time division multiplexing.

**TDMA**   Time division multiple access.

**TE**   Terminal equipment.

**Throughput**   The speed at which data travels through a network.

**UPnP**   Universal plug 'n' play is a higher-layer protocol stack that aims to enable discovery and control of home-networked devices and services.

**Upstream**   Data flow from an end user to a service or network provider.

**USB**   Universal serial bus is a connection between a PC and peripheral delivering high-speed bidirectional serial data transmission.

**UTP**   An unshielded twisted-pair is a wire that is protected with light plastic versus heavy metal and is therefore prone to interference.

**V.90**   An ITU protocol for dial-up modems that supports 53 Kbps for downloading and 33.6 Kbps for uploading.

**V.92**   An ITU protocol for dial-up modems that supports nominal symmetric 56-Kbps transfer rates.

**VDSL**   Very high bit-rate digital subscriber line is the fastest xDSL technology, able to support up to 13-Mbps symmetric traffic or up to 52-Mbps downstream and up to 2.3-Mbps upstream traffic over a single copper-pair wire over a range from 1,000 ft (~0.3 km) to 4,500 ft (~1.3 km).

**VESA**   The Video Electronics Standards Association is a home-networking committee developing an interoperability standard known as the Versatile Home Network, or VHN, standard.

**VFIR**   Very fast infrared.

**VHN**   Versatile Home Network is a standard that defines a digital in-home intranet that connects previously incompatible home components or cluster networks.

**VoD**   Video on demand.

**VoIP**   Voice over IP is a service that enables voice transmission in digital packets over the Internet.

**Wi-Fi**   Wireless-fidelity is a designation by the Wireless Ethernet Compatibility Alliance (WECA) that an 802.11b wireless network component should meet the compatibility standard set for interoperability with other 802.11b products.

**WiMedia**   An open industry forum aiming to develop "standards-based specifications for connecting personal area, wireless multimedia devices." WiMedia promotes the IEEE 802.15.3 standard.

**WLL**   Wireless local loop.

**WRS**   A wireless relay station is a DECT fixed station that extends the coverage of base stations.

**X10**   A technology providing up to 100 Kbps over the existing home power-line wiring tree by exchanging coded low-voltage signals superimposed over the 110V ac current.

# About the Author

Theodore B. Zahariadis received his Ph.D. in electrical and computer engineering from the National Technical University of Athens, Greece, and his Dipl.-Ing. in computer engineering from the University of Patras, Greece. Since 1997, he has been with Lucent Technologies, first as technical consultant to ACT, Bell Laboratories, New Jersey, then as technical manager of Ellemedia Technologies (an affiliate of Bell Laboratories of Advanced Technologies EMEA), Athens, Greece, and since November 2001 as a technical director. At present, he is leading R&D into end-to-end interactive multimedia services. He is responsible for the development of QoS multimedia service platforms, home-network control and management, and embedded systems and wireless protocol implementation. His research interests include broadband wireline, wireless, and mobile communications; interactive service deployment over IP networks; embedded systems; and RGs. He has served as a reviewer for many conferences and magazines and as a guest editor for *ACM/SIGMOBILE MC2R, IEEE Wireless Communications, IEEE Network Magazine*, and *Kluwer Telecom Systems Journal*.

# Index

## Recent Titles in the Artech House Telecommunications Library

Vinton G. Cerf, Senior Series Editor

*SIP: Understanding the Session Initiation Protocol*, Alan B. Johnston

*Smart Card Security and Applications, Second Edition*, Mike Hendry

*SNMP-Based ATM Network Management*, Heng Pan

*Spectrum Wars: The Policy and Technology Debate*, Jennifer A. Manner

*Strategic Management in Telecommunications*, James K. Shaw

*Strategies for Success in the New Telecommunications Marketplace*, Karen G. Strouse

*Successful Business Strategies Using Telecommunications Services*, Martin F. Bartholomew

*Telecommunications Cost Management*, S. C. Strother

*Telecommunications Department Management*, Robert A. Gable

*Telecommunications Deregulation and the Information Economy, Second Edition*, James K. Shaw

*Telecommunications Technology Handbook, Second Edition*, Daniel Minoli

*Telemetry Systems Engineering*, Frank Carden, Russell Jedlicka, and Robert Henry

*Telephone Switching Systems*, Richard A. Thompson

*Understanding Modern Telecommunications and the Information Superhighway*, John G. Nellist and Elliott M. Gilbert

*Understanding Networking Technology: Concepts, Terms, and Trends, Second Edition*, Mark Norris

*Videoconferencing and Videotelephony: Technology and Standards, Second Edition*, Richard Schaphorst

*Visual Telephony*, Edward A. Daly and Kathleen J. Hansell

*Wide-Area Data Network Performance Engineering*, Robert G. Cole and Ravi Ramaswamy

*Winning Telco Customers Using Marketing Databases*, Rob Mattison

*WLANs and WPANs towards 4G Wireless*, Ramjee Prasad and Luis Muñoz

*World-Class Telecommunications Service Development*, Ellen P. Ward

For further information on these and other Artech House titles, including previously considered out-of-print books now available through our In-Print-Forever® (IPF®) program, contact:

| | |
|---|---|
| Artech House | Artech House |
| 685 Canton Street | 46 Gillingham Street |
| Norwood, MA 02062 | London SW1V 1AH UK |
| Phone: 781-769-9750 | Phone: +44 (0)20 7596-8750 |
| Fax: 781-769-6334 | Fax: +44 (0)20 7630-0166 |
| e-mail: artech@artechhouse.com | e-mail: artech-uk@artechhouse.com |

Find us on the World Wide Web at:
www.artechhouse.com